中国古代建筑精粹
The Essence of Ancient Chinese Architecture

民间古堡
Vernacular Castles

张 斌　周晓冬　杨北帆　著
杨 彤（英）　吴 丹　译

中国古代建筑精粹
The Essence of Ancient Chinese Architecture

民间古堡
Vernacular Castles

张 斌　周晓冬　杨北帆　著
杨 彤（英）　吴 丹　译

中国建筑工业出版社

图书在版编目（CIP）数据

中国古代建筑精粹 民间古堡/张斌，周晓冬，杨北帆著.
北京：中国建筑工业出版社，2012.9
ISBN 978-7-112-14535-5

Ⅰ.①中… Ⅱ.①张…②周…③杨… Ⅲ.①古建筑－城堡－介绍－中国 Ⅳ.①TU-092.2

中国版本图书馆CIP数据核字（2012）第173538号

责任编辑：孙立波　程素荣
责任设计：赵明霞
责任校对：张　颖　王雪竹

中国古代建筑精粹
民间古堡
张　斌　周晓冬　杨北帆　著
杨　彤（英）　吴　丹　译
*
中国建筑工业出版社出版、发行（北京西郊百万庄）
各地新华书店、建筑书店经销
北京嘉泰利德公司制版
北京画中画印刷有限公司印刷
*
开本：880×1230毫米　1/32　印张：9　字数：300千字
2012年10月第一版　2012年10月第一次印刷
定价：68.00元
ISBN 978-7-112-14535-5
　　　（22583）

版权所有　翻印必究
如有印装质量问题，可寄本社退换
（邮政编码　100037）

前言

对于喜欢古堡这种浪漫美丽的建筑形式的人而言，对中国古堡可能会有陌生感，因为与欧洲现存 10 万余座古堡、印度每一座古城几乎都有一座大古堡的情况相比，中国的古堡存量实在少了些，地位更不显著，以致很多人会认为中国根本没有古堡。中国的古建筑专业界在划分古建筑类型时不会列出城堡一项，即使现存有一定数量的民间寨堡建筑，但中国的民间建筑统统被归入"民居"一个类别。

事实上，在建筑的发展历史中，城堡是一种重要的建筑类型，它不仅影响了各类建筑的发展，也影响了人类社会、文化的发展。时至今日，作为人类文明的结晶、作为著名历史人物或事件的纪念物、更作为建筑杰作，世界上已经有很多古堡以城堡及要塞这样的建筑类型成为联合国教科文组织认定的世界文化遗产，而城堡及要塞在许多国家的建筑界早已经是独立的建筑类别，对应着设有专门的学术机构。

中国的古堡因为战争破坏、与大一统意识冲突而遭各代王朝清理、文物保护意识曾经淡漠等原因，现存数量确实不多，但它们依然精彩，不应该被忽视遗忘，而且正因为数量少，它们才更显珍贵。特别是，这些古堡，尤其是民间古堡，能在中国古代特殊的历史文化环境中保存下来，实属不易。而且，它们都记录下一段特殊的历史文化信息，不仅丰富了中国古建筑的类型，也丰富了中国的历史文化内容。所以，研究中国民间古堡，一方面是研究建筑，从中获得建筑学的启发；另一方面，还可以得到大量的历史文化信息，了解社会与建筑的互动关系，从而为中国民间对中华文化丰富性的特殊贡献而感动，为太多古堡现正处在风雨飘摇之中而担忧。

Preface

People who appreciate the beauty and romance of the architectural style of ancient castles might be strangers to such in China, because compared to the more than 100,000 castles throughout Europe and virtually one in every ancient Indian city, China is only home to few ancient castles of no significant standing; therefore, it is easy to think ancient castles do not exist in China. While categorizing ancient architecture, the Chinese ancient architecture community would not have a section for vernacular castle, but would place the small number of such in the much broader category of "vernacular dwelling."

In truth, in the history of architecture development, castles are an important architectural style. It influenced not only development of various styles but also that of society and culture. To this day, many ancient castles around the world have already been named World Heritage Sites by UNESCO as examples of refined civilizations, souvenirs of famous historical figures and events, and architectural masterpieces under its own category of castles and forts. Furthermore the study of ancient fortresses has expanded into its own academic field resulting in the foundation of institutions dedicated to it in many countries.

Because the ancient castles in China have been damaged during war times, destroyed by various dynasties due to its contradiction to the idea of the Grand Unification, and victims of indifference regarding preservation of cultural relics, the number still in existence is indeed small. Regardless, they are wondrous to be hold and should not be forsaken, and since they few and far between, they are more precious for it. For these ancient castles, especially ancient vernacular castles, to survive the eclectic cultural shifts throughout Chinese history is miraculous. They each preserve a piece of cultural history, adding to both ancient architecture styles and the cultural history of China. Therefore, researching Chinese ancient vernacular castles would create a new subfield within architecture, while revealing a cornucopia of cultural knowledge that would shed light on the relationship between society and architecture. These unique contributions to the cultural library are enough to inspire and make apparent the endangered state of these architectural relics.

目 录

前言 /v
第一章 古堡建筑概述 /1

一、古堡建筑的界定 /2
1. 古堡元素无处不在 /2
2. 古堡与古城的区别 /4
3. 欧洲古堡建筑简述 /6
4. 古堡建筑的要素和美感 /22

二、中国古堡建筑的特点 /26
1. 特殊社会情况的产物 /26
2. 现存中国古堡的分布和主要类型 /30
3. 与欧洲古堡建筑的异同 /36
4. 民间古堡的精彩和文化价值 /38
5. 历史沿革概要 /40

山西皇城相府的城门洞内
Inside the City gate of Huangcheng Xiangfu, Shanxi.

Contents

Preface/vi

Chapter I   Summary of Ancient Castle Architecture/1

Section 1   Definition of Ancient Castle Architecture/3
1. Universal Elements of Ancient Castles/3
2. Differences Between Ancient Castles and Ancient Cities/5
3. A Brief Narration on European Ancient Castle Architecture/7
4. Elements and Beauty of Ancient Castles/23

Section 2   Distinguishing Traits of Chinese Ancient Castle Architecture/27
1. The Products of Unique Societal Contexts/27
2. Main Types and Distribution of Existing Chinese Castles/31
3. The Similarities and Differences From European Castle Architecture/37
4. The Marvel and Cultural Value of Vernacular Castles/39
5. Chronological Summary of History/41

第二章　黄河流域的民间古堡 /44

一、历史文化背景 /45
1. 农牧交替、胡汉混居 /45
2. 改朝换代、绝境求生 /48

二、山西的特殊性 /48
1. 最丰富的半农半牧区 /48
2. 晋商 /50
3. 农牧商多元文化下的山西古建筑、古堡 /54
4. 活化石现象 /56

三、沁河流域民间古堡 /58
1. 沁水县窦庄 /58
2. 沁水县郭壁 /60
3. 阳城县砥泊城 /66
4. 阳城县郭峪 /70
5. 阳城县"皇城相府" /74
6. 沁水县湘峪堡 /84
7. 沁水县柳家堡 /92
8. 阳城县寨山古堡 /94
9. 泽州县"玫瑰堂"古堡 /96
10. 河南省博爱县寨卜昌村 /98

四、汾河流域民间古堡 /101
1. 平遥县的堡村 /102
2. 介休市张壁 /104
3. 灵石市恒贞堡 /109
4. 灵石市梁家堡 /112

Chapter II　Vernacular Castles of the Yellow River Basin/44

Section 1　The Historical and Cultural Background/47
1. Agricultural/pastoral region, Hu and Han mixed living/47
2. Regime change, desperation to survive/49

Section 2　Particularity of Shanxi Province/49
1. The most plentiful farming/pastoral areas/49
2. Shanxi Merchants/51
3. The multicultural excellency: ancient architecture and castles in Shanxi/55
4. A living fossil phenomenon/57

Section 3　Vernacular Castle in Qinhe River Basin/59
1. Douzhuang village in Qinshui County /59
2. Guobi village in Qinshui County /63
3. Diji town in Yangcheng County/67
4. Guoyu village in Yangcheng County/71
5. Huangcheng Xiangfu in Yangcheng County/75
6. Xiangyubu Castle in Qinshui County/85
7. Liujiabu Castle in Qinshui County/93
8. Zhaishan Castle in Yangcheng County/95
9. "Rose Hall" Castle in Zezhou County/97
10. Zhaibuchang Village in Bo'ai County, Henan Province/99

Section 4　Vernacular Castle in the Fenhe River's Basin /101
1. Castle villages in Pingyao County/103
2. Zhangbi Castle in Jiexiu City/105
3. Hengzhenbu Castle in Lingshi City/110
4. Liangjiabu Castle in Lingshi city /113

五、陕、宁、鲁等省、自治区的民间古堡 /116

1. 陕西横山县波罗堡 /116

2. 宁夏吴忠市董家堡 /118

3. 山东惠民县魏家堡 /121

4. 山东肥城市古堡群 /124

第三章 闽粤赣3省的民间古堡 /126

一、历史文化背景 /127

1. 人口大迁徙 /127

2. 山海之间 /128

3. 宗法与风水 /130

二、赣南四角楼 /132

1. 龙南县关西新围 /134

2. 乌石村 /138

3. 燕翼围 /138

三、粤北四角楼 /140

1. 和平县林寨四角楼 /140

2. 始兴县的大碉楼 /144

3. 司前镇四角楼 /148

4. 隘子镇满堂围 /150

5. 翁源县的多角楼 /152

四、沿海古堡 /154

1. 闽东南的寨中楼 /155

2. 潮州堡寨 /160

Section 5 Vernacular Castles in Shaanxi, Ningxia and Shandong/117

1. Buoluobu Castle: Hengshan County, Shaanxi Province/117

2. Dong Family Castle in Wuzhong City, Ningxia/119

3. Wei Family Castle in Huimin County, Shandong Province/123

4. Fei City Castle Group in Shandong Province/125

Chapter III Vernacular Castles in the Three Provinces of Fujian, Guangdong and Jiangxi/126

Section 1 The Historical and Cultural Background/129

1. The Great Migration/129

2. Amongst mountains and oceans/131

3. Patriarchal system/Patriarchy, and *Feng Shui*/131

Section 2 Sijiaolou in the South of Jiangxi /133

1. Longnan Guanxi Xinwei/135

2. Wushi Village/139

3. Yanyiwei Castle/139

Section 3 Northern Guangdong, Sijiaolou (Castle with four-corner-tower)/141

1. Heping County, Linzhai Town, Sijiaolou/141

2. Diaolou in Shixing County/145

3. Siqian Town Sijiaolou/149

4. Mantangwei in Aizi Town /151

5. Wengyuan County's Multi-corner towers /153

Section 4 The Coastal Castle/154

1. Fujian Southeast Keep Tower/157

2. Chaozhou Castle/161

ix

第四章 闽南粤东的圆土楼 /163

一、对圆土楼来源的探讨 /164
1. 对现有解释的疑问 /164
2. 圆形的神圣性与原始性 /166
3. 关注井 /168
4. 漳州潮州交界处的历史文化特征 /175
5. 凤凰山土楼的特征 /179
6. 大胆假设与小心求证 /184

二、粤东土楼 /184
1. 饶平县道韵楼 /184
2. 上饶镇镇福楼 /186
3. 上善镇南华楼 /189

三、福建圆土楼 /190
1. 漳浦县锦江楼 /190
2. 平和县椭圆形砖楼 /192
3. 绳武楼和丰作厥宁楼 /196
4. 永定县承启楼 /199
5. 洪坑村振成楼 /202

Chapter IV Round Earth Buildings in South of Fujian and East of Guangdong/163
Section 1 Investigate/Trace the Source of the Round Earth Buildings/165
1. Doubt on the existing interpretations/165
2. The sanctity and the primitives of round shape/167
3. Focusing on the well/171
4. Historical and cultural characteristics in the border region of Zhangzhou and Chaozhou/176
5. Characteristics of earth buildings in the Mountain Phoenix/181
6. Bold assumptions and careful verification/185

Section 2 Earth Building in Eastern Guangdong/185
1. Daoyunlou in Raoping county/187
2. Zhenfulou in Shangrao town/188
3. Nanhualou in Shangshan town/189

Section 3 Round Earth Buildings in Fujian Province/191
1. Jinjianglou in Zhangpu County /191
2. Oval brick building in Pinghe county/193
3. Shengwulou and Fengzuo Juening Building/198
4. Chengqilou in Yongding County /201
5. Zhenchenglou in Hongkeng Village/203

江西关西新围角楼枪炮孔外
Look out from the gun hole of the corner tower in Guanxi Xinwei, Jiangxi Province.

第五章 文化融合的大古堡 /207

一、古堡文化的集成 /208
1. 南北、土客、华夷文化的融合 /208
2. 古堡设计的成熟 /210

二、闽中大堡 /210
1. 三明市莘口镇松庆堡 /210
2. 水美 3 堡 /210
3. 永安市安贞堡 /217

三、闽西南方土楼和府第式土楼 /222
1. 永定县抚市镇实善楼 /222
2. 五福楼 /224
3. 永隆昌楼 /226
4. 洪坑村福裕楼 /228

Chapter V Culturally Integrated Castles/207
Section 1 Integration of Castles into Culture/209
1. Cultural Integration between: North and South, the natives and the Hakkas, the Han people and the people of various minorities/209
2. Maturation of Castle Design/211

Section 2 Great Castles in the Middle of Fujian/211
1. Songqingbu Castle in Xinkou Town, Sanming City/211
2. Three Castles in Shuimei village/211
3. Anzhenbu Castle in Yong'an City/219

Section 3 Earth Buildings in Southwestern Fujian and the Mansion-Style Earth Buildings/223
1. Shishanlou, Fushi Town, Yongding County /223
2. Wufulou Building /225
3. Yonglongchang Building /227
4. Fuyulou, Hongkeng Village /229

四、广东大围楼 /230

1. 大埔县泰安楼 /230
2. 东源县康禾镇仙坑村 /232
3. 乐村石楼 /234
4. 三顺楼 /239
5. 叶潭镇山下八角楼 /244
6. 紫金县德先楼 /246
7. 新丰县儒林第 /251
8. 增城市邓村石屋 /253
9. 深圳龙岗鹤湖新居和龙田世居 /254
10. 深圳东莞碉楼 /257
11. 开平碉楼 /260

第六章　其他古堡 /263

一、军事转民用的古堡 /264
1. 江西省寻乌县羊角水古堡 /264
2. 贵州省安顺市屯堡 /268

二、湖北省南漳县古堡 /271

三、羌藏地区的民间古堡 /273

Section 4　Guangdong Grand Enclosed House/231

1. Tai'anlou, Dapu County/231
2. Xiankeng Village, Kanghe Town, Dongyuan County /233
3. Stone Building, Lecun Village/235
4. Sanshunlou Castle/241
5. Eight-corner-tower Castle in Shanxia Village, Yietan Town/245
6. Dexian Castle, Zijin County/248
7. Rulindi, Xinfeng County/252
8. Stone Castle, Dengcun Village, Zengcheng City /253
9. Hehu Xinju and Longtian Shiju, Longgang, Shenzhen/255
10. Diaolou in Dongguan, Shenzhen City/259
11. Diaolou in Kaiping County /261

Chapter VI　Other Castles/263

Section 1　Military-turned-civilian Castles/265
1. Yangjiaoshui Castle, Xunwu County, Jiangxi Province /265
2. Official's Military Fortress, Anshun City, Guizhou Province/269

Section 2　Nanzhang County Castles, Hubei Province/272

Section 3　Qiang and Tibetan Vernacular Castles/274

江西关西新围角楼枪炮孔外
Look out from the gun hole of the corner tower in Guanxi Xinwei, Jiangxi Province.

第五章　文化融合的大古堡 /207

一、古堡文化的集成 /208
1. 南北、土客、华夷文化的融合 /208
2. 古堡设计的成熟 /210

二、闽中大堡 /210
1. 三明市莘口镇松庆堡 /210
2. 水美 3 堡 /210
3. 永安市安贞堡 /217

三、闽西南方土楼和府第式土楼 /222
1. 永定县抚市镇实善楼 /222
2. 五福楼 /224
3. 永隆昌楼 /226
4. 洪坑村福裕楼 /228

Chapter V　Culturally Integrated Castles/207
Section 1　Integration of Castles into Culture/209
1. Cultural Integration between: North and South, the natives and the Hakkas, the Han people and the people of various minorities/209
2. Maturation of Castle Design/211

Section 2　Great Castles in the Middle of Fujian/211
1. Songqingbu Castle in Xinkou Town, Sanming City/211
2. Three Castles in Shuimei village/211
3. Anzhenbu Castle in Yong'an City/219

Section 3　Earth Buildings in Southwestern Fujian and the Mansion-Style Earth Buildings/223
1. Shishanlou, Fushi Town, Yongding County /223
2. Wufulou Building /225
3. Yonglongchang Building /227
4. Fuyulou, Hongkeng Village /229

四、广东大围楼 /230

1. 大埔县泰安楼 /230
2. 东源县康禾镇仙坑村 /232
3. 乐村石楼 /234
4. 三顺楼 /239
5. 叶潭镇山下八角楼 /244
6. 紫金县德先楼 /246
7. 新丰县儒林第 /251
8. 增城市邓村石屋 /253
9. 深圳龙岗鹤湖新居和龙田世居 /254
10. 深圳东莞碉楼 /257
11. 开平碉楼 /260

第六章　其他古堡 /263

一、军事转民用的古堡 /264

1. 江西省寻乌县羊角水古堡 /264
2. 贵州省安顺市屯堡 /268

二、湖北省南漳县古堡 /271

三、羌藏地区的民间古堡 /273

Section 4　Guangdong Grand Enclosed House/231

1. Tai'anlou, Dapu County/231
2. Xiankeng Village, Kanghe Town, Dongyuan County /233
3. Stone Building, Lecun Village/235
4. Sanshunlou Castle/241
5. Eight-corner-tower Castle in Shanxia Village, Yietan Town/245
6. Dexian Castle, Zijin County/248
7. Rulindi, Xinfeng County/252
8. Stone Castle, Dengcun Village, Zengcheng City /253
9. Hehu Xinju and Longtian Shiju, Longgang, Shenzhen/255
10. Diaolou in Dongguan, Shenzhen City/259
11. Diaolou in Kaiping County /261

Chapter VI　Other Castles/263

Section 1　Military-turned-civilian Castles/265

1. Yangjiaoshui Castle, Xunwu County, Jiangxi Province /265
2. Official's Military Fortress, Anshun City, Guizhou Province/269

Section 2　Nanzhang County Castles, Hubei Province/272

Section 3　Qiang and Tibetan Vernacular Castles/274

xii

第一章 古堡建筑概述

Chapter I Summary of Ancient Castle Architecture

一、古堡建筑的界定

1. 古堡元素无处不在

"古堡"是现代人对古代城堡、要塞等建筑形式的称谓。几十年前，中国人也许觉得自己离古堡很远，然而实际上，一些耳熟能详的词汇就是古堡，如前苏联、现俄罗斯克里姆林宫的"克里姆林"、古巴卡斯特罗主席的"卡斯特罗"，原意都是古堡。

中国开放以后，一些欧美词汇也进入中国，如美国华尔街。后来，人们深入地了解到，"华尔"是指纽约还是荷兰人的新阿姆斯特丹时，荷兰人所建城堡要塞的一道墙体。以为现代人打造梦想世界来运营文化娱乐业的美国迪斯尼公司的标志物除了卡通形象，就是《灰姑娘》中的城堡，其城堡形象主要来自德国的新天鹅堡，而新天鹅堡的设计灵感来自德国爱森纳赫的瓦尔特堡，瓦尔特堡又曾经是宗教改革运动中马丁·路德的避难所。另一座时常出现在画册上的瑞士西庸古堡除了美，还因为英国诗人拜伦在那里写下《西庸的囚徒》而著名，使其成为浪漫主义的圣地。

克里姆林宫
Kremlin Palace, Moscow, Russia.

Section 1 Definition of Ancient Castle Architecture

1. Universal Elements of Ancient Castles

Ancient castles is thusly termed by modern scholars when referring to castles and fortresses of ancient times. Decades ago, the idea of ancient castles was far-fetched in the minds of Chinese people, but in reality, many familiar names mean ancient castles such as the Moscow "Kremlin" and Fidel "Castro."

Soon after the Economic Reforms, western phrases like Wall Street made their way into Chinese language. The name "Wall" comes from pre-colonial times when the Dutch occupied the now southern portion of the island of Manhattan when the northern wall of said fortification lied along what now houses the epicenter of world finance. Simultaneously a symbol for a magical world and an entertainment industry superpower, the Disney logo of *Sleeping Beauty* Castle is modeled after Neuschwanstein Castle in Schwangau, Germany, the design of which was inspired by Wartburg Castle in Eisenach, Germany which was the hiding place of Martin Luther after his excommunication from the Catholic Church. Château de Chillon has been incorporated into many works of art not only because of its beauty but also a result of Lord Byron's poem, *"The Prisoner of Chillon,"* making it the epitome of romanticism.

德国瓦尔特堡
Wartburg Castle, Eisenach, Germnay.

瑞士西庸堡
Chillon Castle, Geneva, Switzerland.

不知不觉中，中国许多原为××堡、××寨、××壁的地名中的最后一个字消失了，或改成了"村"字等，但还是有许多地方的地名中仍然保留着"堡"、"寨"、"壁"等字，历史记忆告诉人们，那些地方原来应该都有古堡。随着乡土观念的回归和正面化，人们不再认为古堡是象征落后的地方，许多保留下来的寨堡成为文化财富。

2. 古堡与古城的区别

古堡最鲜明的形象标志是高大的防御性城墙，这样，古堡就容易与有城墙的古城混淆，区别二者，主要在三个方面。

首先是大小，除了"堡"字，汉字中指古堡的主要还有"坞"、"壁"、"寨"等字，这几个字在古文里都有小城的意思。城的意思是容民的设施，古城都有城墙，用以护民。

其次，城市与城堡的区别还在于城市突出"市"的属性，即商业内容，城堡突出堡，即防御功能。城堡可以作为市场的看护者，但很少容纳市场。

山西湘峪堡　Xiangyu bu Castle, Shanxi.

Section 1   Definition of Ancient Castle Architecture

1. Universal Elements of Ancient Castles

Ancient castles is thusly termed by modern scholars when referring to castles and fortresses of ancient times. Decades ago, the idea of ancient castles was far-fetched in the minds of Chinese people, but in reality, many familiar names mean ancient castles such as the Moscow "Kremlin" and Fidel "Castro."

Soon after the Economic Reforms, western phrases like Wall Street made their way into Chinese language. The name "Wall" comes from pre-colonial times when the Dutch occupied the now southern portion of the island of Manhattan when the northern wall of said fortification lied along what now houses the epicenter of world finance. Simultaneously a symbol for a magical world and an entertainment industry superpower, the Disney logo of *Sleeping Beauty* Castle is modeled after Neuschwanstein Castle in Schwangau, Germany, the design of which was inspired by Wartburg Castle in Eisenach, Germany which was the hiding place of Martin Luther after his excommunication from the Catholic Church. Château de Chillon has been incorporated into many works of art not only because of its beauty but also a result of Lord Byron's poem, *"The Prisoner of Chillon,"* making it the epitome of romanticism.

德国瓦尔特堡
Wartburg Castle, Eisenach, Germnay.

瑞士西庸堡
Chillon Castle, Geneva, Switzerland.

不知不觉中，中国许多原为××堡、××寨、××壁的地名中的最后一个字消失了，或改成了"村"字等，但还是有许多地方的地名中仍然保留着"堡"、"寨"、"壁"等字，历史记忆告诉人们，那些地方原来应该都有古堡。随着乡土观念的回归和正面化，人们不再认为古堡是象征落后的地方，许多保留下来的寨堡成为文化财富。

2. 古堡与古城的区别

古堡最鲜明的形象标志是高大的防御性城墙，这样，古堡就容易与有城墙的古城混淆，区别二者，主要在三个方面。

首先是大小，除了"堡"字，汉字中指古堡的主要还有"坞"、"壁"、"寨"等字，这几个字在古文里都有小城的意思。城的意思是容民的设施，古城都有城墙，用以护民。

其次，城市与城堡的区别还在于城市突出"市"的属性，即商业内容，城堡突出堡，即防御功能。城堡可以作为市场的看护者，但很少容纳市场。

山西湘峪堡　Xiangyu bu Castle, Shanxi.

长城关城嘉峪关
Jiayuguan Pass, the Great Wall.

明代军事戍堡山西的广武城
Guangwu City (Military Fortress in Ming Dynasty), Shanxi.

Little by little, names of some settlements that used to contain words such as "castle," "fort," and "keep" were replaced by "village" and other words while other retained their original name. History tells us that places named "castle," "fort," and "keep" should have such structures within its boundaries. The reemergence of positive provincialism allowed people to no longer view castles and forts as signs of underdevelopment and many preserved castles and forts became cultural treasures.

2. Differences Between Ancient Castles and Ancient Cities

The most distinctive feature of ancient castles is ramparts built for defense, which is easily confused with ancient cities that also have a defensive wall surrounding them. Three aspects differentiate the two types of ancient structures.

Firstly, ancient castles are smaller than ancient cities. In Chinese words synonymous to castle, such like Wu, Bi, Zhai, that are often used in its place all have an implied meaning of "small city," a city being shelter for its citizens. Ancient cities also have walls surrounding them to protect its citizens.

Secondly, cities differ from castles in function. Cities are centers of trade and commerce while castles are mainly built for protection. A castle can overlook a marketplace but could rarely house one.

山西陈氏家族古堡"皇城相府"　Chen family's Castle – Huangcheng Xiangfu, Shanxi.

最后,城市是公共场所,城堡是小团体或私人场所,只在紧急情况时可能成为公共场所。

一般来讲,常规的城市与城堡容易区别,在中国,不容易区别的主要是古代的军屯、关城等政府城堡,以及各种设防村镇,本书主要记述民间古堡,但民间古堡的形式与军屯等有互动关系,军屯等也有可能慢慢转变成为民间古堡。

多数古堡位于远离城市的郊野,但也有一些古堡是城市防御系统的一部分,位于城市中间或城市的一角,并与城墙联系在一起。还有的城市是从一座城堡或围绕着一座城堡发展起来的,如莫斯科就是围绕克里姆林宫形成的。

3. 欧洲古堡建筑简述

历史学者普遍认为,古堡建筑与封建制度是对应的,中国古代社会并不是典型的封建社会,所以古堡建筑的发展才缺乏连贯性,类型、数量才有限。欧洲中世纪社会是最典型的封建社会,所以欧洲古堡的发展脉络最清晰、类型多、存量大。我们可以通过简单梳理一下欧洲古堡的情况,来对我们认识中国古堡提供帮助。

希腊罗德港口古堡　Ancient Castle, Rhodes harbour, Greece.

Lastly, a city is a public forum while castles are accessible by a select few and would only be open to the public in an emergency.

Generally, typical cities and castles are easy to differentiate. More difficult to differentiate are official military castles, military outposts, and villages with defensive capabilities from castles. This book primarily focuses on vernacular castles, but these other types of castles and vernacular castles can be easily interconverted.

Castles are often a part of the defensive capabilities of a city, located in the center or a corner of the city, connecting the ramparts. Other cities were established near or around castles, the same way Moscow developed around the Kremlin.

3. A Brief Narration on European Ancient Castle Architecture

Historians believe feudalism provided a basis and necessity of ancient castle architecture. Because ancient China was not a typital feudal society, the development of ancient castle architecture lacks consistency and varying types of castles and their abundance are limited. Medieval Europe was the most basic feudal society, and the progression of feudalism as well as the development of castle architecture can be clearly traced back to that it as a result of the abundance in quantity and variety of castles throughout Europe. We can, through a general understanding of the history of European castles, better acquaint ourselves with Chinese ancient castles.

虽然欧洲人对古堡的准确界定是中世纪封建领主所建的，用以保护、争夺领地的武装建筑，但更早的如罗马帝国的军事要塞等并不能被完全排除在城堡建筑类型之外，事实上，二者在英语中的名称非常接近(castle/castra)。早期的许多城堡或本身是由古罗马要塞发展而来的，如法国的卡尔卡松城堡；或是在建造时参照了古罗马要塞的形式；或是结合了古罗马建筑的结构，如法国阿尔勒人曾经居住在古罗马竞技场中，用竞技场的外墙当防卫城墙。而更早的古希腊卫城建筑、伊特鲁利亚人在亚平宁半岛上筑的小山城等，实际上也是一种城堡。

至于东罗马帝国（拜占庭帝国）的城堡，应该是古罗马要塞和西亚、中东地区古要塞建筑的集成之作。拜占庭统治意大利和十字军东征时期，是城堡建筑技术由于交流广泛而迅速提高的一个时期。

希腊罗德岛的罗德老城是一座包含城堡的武装城市，由在十字军东征时期成立的圣约翰医院骑士团建立。之前，它是著名的古希腊城邦城市，其防御体系让进攻的马其顿人丢下辎重退却，从而成就古代世界七大奇迹之一的太阳神灯塔。在十字军完全丢失中东后，医院骑士团来到罗德岛抵御奥斯曼人，他们所建的坚固要塞、城堡曾经让奥斯曼人长时间一筹莫展。同时，要塞、城堡又非常美丽，有雄浑造型的高大石墙散发出动人的质感，适度而精美的石雕装饰带来高贵的气质。如果说被城墙围绕的老城是城市，不是城堡的话，那么骑士团的大团长官邸就完全是一座城堡，尽管人们习惯称它为宫殿。骑士团在罗德岛另一座古希腊城市林都斯的城堡则是结合古希腊的卫城建造的。

德国特里尔古罗马城门，其自身的体量就是一座古堡
The Roman city gate, Trier, Germany (It can be regarded as a castle by its sheer volume and scale).

法国阿尔勒古罗马竞技场曾经作为城堡使用
The Roman Arena, Arles, France (It had used as a castle).

Europeans understand castles to be defined by the way in which they were built during medieval times for both protection and offensive ability, but the military fortifications of the Roman Empire cannot be excluded from the category. In fact, "castle" is derived from the Latin word for the Roman fortifications, "castrum". Most of the earliest castles are original Roman military fortifications like Cité de Carcassonne in France, were built in the style of such Roman fortifications, or were structured like ancient Roman architecture such as the outer walls of Arles, France. The acropolises of ancient Greece and the Etruscan cities on the Apennine Peninsula are also castles.

Castles of the Byzantine Empire exhibit architectural elements from castles of ancient Rome, western Asia, and the Middle East. The increased flow of information during the Crusades allowed Byzantine architects to develop more advanced techniques in that time period.

The Colossus of Rhodes, one of the Seven Wonders of the Ancient World, was erected in celebration of the unsuccessful Siege of Rhodes. the Knights Hospitaller which established after the First Crusades, settled and built castles in the city of Rhodes, Greece. These castles and defensive structures helped the Hospitallers in resisting siege by Ottoman troops in the late 1400s. In addition to their practical uses, these castles are surrounded by forceful

罗德岛林都斯利用古希腊卫城建的城堡
Acropolis of Lindos,an ancient castle, Rhodes, Greece.

罗德医院骑士团大团长官邸入口
The entrance of the Palace of the head of Rhodes Knights' Hospital, Greece.

意大利那不勒斯的蛋堡，它由古希腊殖民者开基，哥特人曾经在此囚禁最后一位西罗睿帝国皇帝，后来法国人、西班牙人等不断续建
Castle dell'Ovo Naples, Italy, the core of architecture was initialized by the ancient Greek colonists. Goths had imprisoned the last emperor of the Western Roman Empire here, and later the French and Spanish continued the construction.

出于安全的考虑，中世纪早期的宫殿，乃至许多教堂、修道院建筑都是城堡形式，如罗德岛附近的帕特莫斯岛上的基督教圣地圣约翰修道院。

法国阿维尼翁的教皇宫殿也是城堡形式，挟持教皇的法国国王腓力四世的城堡与教皇宫殿隔罗讷河相望，造型与罗德岛骑士城堡很接近，带雉堞的城墙和高大的塔楼是城堡的标志性构造。

作为海盗维京人的一支、作为最早与南欧拜占庭及阿拉伯人接触的西北欧人、作为十字军的积极参与者，诺曼人统治过南意大利、法国诺曼底及安茹、巴勒斯坦及中东数地，后来又征服了英格兰及威尔士。诺曼人的战斗力一部分来自于他们是筑城堡的能手，英国除了古罗马人留下的要塞，最早的古堡就是征服者威廉攻入英格兰后建的，如伦敦塔和温莎堡的基础，其中伦敦塔原来是古罗马要塞。法国的许多古堡也具有诺曼特征：有简单的几何性，突出大型圆

法国阿维尼翁教皇宫殿
Palace of the Popes, Avignon, France.

教皇宫殿对面的法王城堡
The French King's Castle, opposite to the Palace of the Popes.

蒙特堡
Castel del Monte, Abruzzo, Italy.

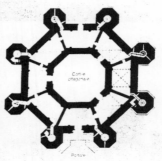

蒙特堡平面图
The plan of the Castel del Monte.

stonewalls that radiate raw power and decorated with modest yet stunning stone sculptures, giving the castle or fortress and air of elegance and nobility. If one was to say that the walls surrounding the ancient city encompasses the whole city, and not a castle, then the living quarters of Grand Master of the Knights Hospitaller is definitely a castle, even though people often refer to it as a palace. The Knights later built a fortress upon the remnants of an older Byzantine fortification on the acropolis in Lindos.

Due to safety, early medieval palaces, churches, and monasteries were structured like castles, for example, the Christian holy site Monastery of Saint John, which resides on the nearby island of Patmos.

The Palace of Popes in Avignon, France, which looks upon the palace of King Philip IV across the Rhone, shows characteristics of a castle and was fashioned after castles built by the Knights of Rhode. Walls mounted with battlements and towers are constructs indicative of a castle.

The Vikings, or Norsemen, once occupied southern Italy, French Normandy, Anjou, Palestine, and the Middle East, and later, England and Wales. The Vikings' military power is partially due to their mastery of building castles. The earliest castles in England, excluding those built by ancient Romans, such as Windsor Castle and the Tower of London, which was built on the foundation of a Roman fortress, were built by William the Conqueror. Many French castles, such as Château de Vincennes in Paris, which is one of the same type as the Bastille, and those in Rouen, Angers, and Nantes exhibit Norse influence and employ simple geometry, primarily large cylindrical towers.

希腊帕特莫斯岛圣约翰修道院　　St.John Monastery, Patmos Greece.

温莎堡　The Windsor Castle.

塔楼，如鲁昂、昂热、南特城堡和巴黎的文森城堡等，后者与著名的巴士底城堡属于同种类型。

更加有意识地在城堡设计中采用几何学的是神圣罗马帝国皇帝和西西里、耶路撒冷国王腓特烈二世，他主要生活在从诺曼人母亲那里继承的西西里王国，受到阿拉伯文化的影响，熟悉阿拉伯语和领导过第六次十字军东征也使他更加熟悉当时主要由拜占庭人和阿拉伯人掌握的几何学，他在南意大利建造的蒙特堡以八角形为母题，大大小小的八角形之间衔接规则。腓特烈二世的建筑对后来的文艺复兴建筑必然产生影响，其影响力应该还包括诗歌领域，因为但丁曾经誉他为意大利诗歌之父。

虽然同时是德国国王，但腓特烈二世很少管德国事务，更没有在德国建城堡，德国等中东欧地区的城堡在整个中世纪乃至后来的年代里都是一种自然主义式的民间哥特风格，这种哥特风格不同于大教堂、市政厅等建筑采用的那种哥特风格，它更有原始和来自田园的美感，使它们后来最符合浪漫主义和童话世界的审美。

欧洲中世纪的城堡最重视的是防御性，即使它们是国王、主教、贵族的居所，也不过多考虑生活的舒适性、丰富性，这使这些城堡后来多被直接改成监狱。

Holy Roman Emperor Fredrick II, who primarily resided in Sicily, was influenced by Arabic culture, spoke Arabic, and lead the Six Crusade, was familiar with geometry, which at the time was only understood by scholars from the Byzantine Empire and Arab states. He designed Castel del Monte in southern Italy, which consisted of small and large octagons organized in perfect symmetry. Fredrick II influenced the Renaissance not only with his architectural designs, but also in poetry, having been dubbed the father of poetry by Dante.

上2图：德国莱茵河边的古堡　　The two photos above: Castles along Rhine River, Germany.

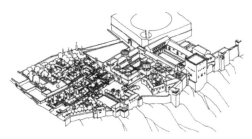

阿尔罕布拉宫鸟瞰图
The bird's eye view of the Alhambra Palace, Granada, Spain.

古堡中的摩尔花园
The Moore Garden in the castle.

古堡的部分建筑被现代人改造为古堡酒店
Part of the castle has been renovated to a modern Castle hotel.

而征服伊比利亚半岛的摩尔人在安达卢西亚的艳阳下不能拒绝美好生活，他们也住在城堡里，但城堡里满是宫殿花园，不是最后避难用的大碉楼，这可能也是摩尔人最终失败的原因之一，但欧洲人会感谢他们留下了美丽古堡。

　　随着战乱减少、随着在火炮发展后防御性不能再都指望城堡、随着文艺复兴运动传播新的生命价值观、也因为看到了摩尔人的城堡生活，欧洲人也开始把他们的城堡改造为花园城堡，城堡的防御性能越来越低，有的根本不设防，完全像是文艺复兴式的别墅，只是它们仍然是古堡的名义和气息。

米兰斯福尔扎古堡
Sforza Castle, Milan, Italy.

法国舍农索古堡
Chenonceau Castle, Chenonceaux, France.

法国昂不瓦斯古堡
Amboise Castle, Loire Valley, France.

Even though Fredrick II was simultaneously King of Germany, he rarely concerned himself with German affairs and even more so did he construct any castles in Germany. Castles of Germany and countries of the Middle East Europe from the medieval times on were built in the vernacular gothic style, different from the gothic style of cathedrals and town halls. These castles' primitive and idyllic beauty became the archetypal romantic settings of cartoon animation.

Castles of medieval Europe were primarily built as a defense mechanism, and, even though they were residences of kings, bishops and other nobility, were not fashioned for luxury or diversity, thus making them readily converted into prisons in later times. Conversely, the Moors in Andalusia could not forego such luxuries, and the castles they lived in were full of gardens instead of keeps or citadels, which partially led to their downfall during the Reconquista. Nevertheless, the charm of these castles continues to be appreciated by Europeans.

With the invention of cannons, castles lost their reinforcing capabilities, as they were not built to withstand cannonballs, and many were converted to mimic the castles the Moors built. Some lost all defensive purposes completely and styled after Renaissance villas, but still maintained the name and air of castles.

Castello Sforzesco in Milan, with large square towers and intimidating walls, could not be deemed indefensible, but when Francis I threatened the castle with mines, the French took Castello Sforzesco. Da Vinci, who painted the ceiling and walls of Sala Delle Asse in Castello Sforzesco, was invited by Francis I to participate in the design of Château de Chambord on the edge of the River Loire. After only three years in France, da Vinci passed away and was buried at the chapel of the nearby Château d'Amboise.

The combination of straddling the River Cher, the terrace gardens, and the rural locale of Château de Chenonceau creates a vista that is unique with classical inspiration to mansions and manors of later generations. Château de Chenonceau was also host to Voltaire, Montesquieu, Rousseau, and other scholars of literature and philosophy who discussed such in salons all over Europe, especially France.

The line between the architectural design of palaces and castles became blurred around that time period, and from it came Château de Chantilly, which was not constrained to the architectural style of either type of structure.

As a solution to the threat of cannons to castles, a new type of defensive structure, forts, came into being. Forts, instead of height, pursue depth within the walls and sections that can each be defended while retreating into the heart of the fort. Ideally, forts have no blind spots along its walls so that every inch of it can be defended, such as a fort that is in the

意大利米兰的斯福尔扎城堡，规则方形，城墙塔楼巨大，防御性不可谓不好，但在法国国王弗朗索瓦一世的地雷战面前也只能开城投降。曾经为斯福尔扎城堡搞过装饰的达·芬奇不久被弗朗索瓦一世请到法国，参与了卢瓦尔河边的香波堡的设计，但他只在法国活了3年，死后葬在香波堡西面的昂不瓦斯皇家城堡里。

昂不瓦斯城堡附近还有舍农索城堡，它基本上是不设防的，它的跨河城堡建筑、水中台地花园、农庄的组合使它成为一座古堡经典，对后世的古堡式府第、庄园设计影响很大。在法国流行沙龙文化时期，舍农索也有著名的沙龙，伏尔泰、孟德斯鸠、卢梭等人都曾经是那里的客人。

此时期的法国王室、贵族建筑在宫苑、城堡类型之间的界限很模糊，很大的自由度使一些精品产生出来，如尚蒂伊城堡，设计与舍农索一样灵活。

为应对火炮对城堡越来越大的威胁，一种新的防御建筑——要塞（fort）逐渐产生。与城堡尽量要高大不同，要塞重点增加纵深和层次，城墙尽量不留侧面防守死角。作为一种曾经深奥的军事技术，对应的要塞技术专家们像达·芬奇一样出名，如法国元帅沃邦。成熟的要塞模式是五角星形堡垒或称棱堡，这种堡垒与一般概念的城堡不同，但许多城堡和要塞是结合在一起的，城堡外围常常增建要塞，而要塞在废弃后，其内部或上面常常建起其他建筑，使其状若古堡。

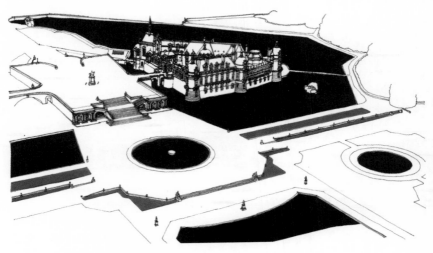

法国尚蒂依古堡　　Chantilly Castle, Chantilly, France.

shape of a five-pointed star, where every wall can be covered by another. Fort engineers who also understand military tactics are famous for their genius, such as Marshal of France, Vauban. The most tested design is forts if the aforementioned five-pointed star shape, nicknamed "star fort." Forts became common additions to the exterior of castles, but once they were no longer needed, they were further remodeled to look like castles.

Site plan of the star fort of Pamplona. It can be seen that like town walls within which a castle sits, forts can also protect a castle, replacing the town wall.

Martin Luther, after starting the Protestant Reformation, hid away in Wartburg Castle while others built or remodeled castles in order to protect their own beliefs. Prisons converted from castles were also used at that time to impose religious opinions on its residents. Later, Henry IV was confined to Vincent Castle in Paris for a time for being a Huguenot and could only assume the throne after renouncing Protestantism. In 1598, he issued the Edict of Nantes, which granted civil rights to Calvinist Protestants and promoted religious tolerance,

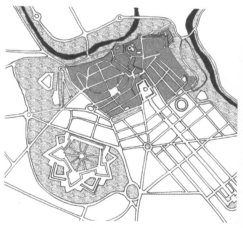

西班牙潘普洛纳的星形要塞平面图。从中也可见要塞与城墙的关系，相当于原来城墙与城堡的关系
Pamplona, Spain's star-shaped fortress plan. The links between the fortress and city walls can also be seen, that equivalent to the relations between the original city walls and castle.

右 2 图：潘普洛纳要塞的细部
The two figures right: Details of the fortress of Pamplona.

马丁·路德发起宗教改革运动后，他自己在瓦尔特城堡中避难，全欧洲陷入宗教纷争的人纷纷新建或加固老城堡，以此来坚守自己的信仰，用古堡改成的监狱也在强迫入狱的人改变信仰。后来的法国国王亨利四世因为信仰新教，被关在巴黎文森城堡中一段时间，他被迫改变信仰才保住性命和实现登基。1598年，他在南特城堡旁边的大教堂里颁布《南特敕令》，推行宗教宽容政策，给予法国新教徒信仰自由和合法权利，而保障就是允许新教徒拥有自己的城堡，即拥有有效自卫权。

然而在1685年，路易十四废除了他爷爷的宽容法令，进攻新教徒的城堡。新教徒最大的要塞——大西洋边的港口城市拉罗谢尔在此之前就遭到攻击，虽然它的要塞坚固，但它还是被攻破。在今天，拉罗谢尔的庞大要塞与港口古堡结合，生成动人的景色。

以路易十四为代表的集权形国王不喜欢住古堡，他住辽阔开敞的凡尔赛宫。他也不喜欢他的臣民包括贵族拥有城堡，特别是设防城堡，认为那是想抵抗他意志的象征。在强化国王权力时期，欧洲各国都有许多城堡被拆毁。

古堡的复兴、特别是古堡文化的复兴与浪漫主义思潮的兴起密切相关。当工商业彻底取代农业成为经济主导产业后，资产阶级和贵族中的自由主义分子更加不满国王的集权专制，法国因此爆发了大革命；英国社会虽然已经由资产阶级主导，但新生成的等级社会仍然令人窒息，所谓上流社会唯利是图又矫揉造作；奥地利和德国还是封建制；欧洲社会急需清新空气。

巴黎文森城堡
Vincent Castle, Paris, France.

上图：法国南特城堡，亨利四世当时住在这里
Above: Nantes Castle, Loire Valley, France (King Henry IV had lived here).

at Château des ducs de Bretagne, effectively ending the religious wars. With their civil rights reinstated, Calvinist Protestants were able to acquire their own castles to effectively protect themselves.

Louis XIII, Henry's son and successor, successfully besieged La Rochelle in response to revolt by some of its Protestant residents. His son, Louis XIV, later revoked the Edict of Nantes, once again refusing Protestants their civil rights. In present say, the La Rochelle harbor is incorporated into the remnants of the castle providing a stunning view of rich history combined with radiant city life.

Louis XIV represented monarchs did not like to live in castles. Instead, he lived in the expansive Palace of Versailles. He did not like his subjects, peasants and nobles alike, living in castles because he saw it as a symbol of subversion. Many castles were demolished while European leaders sought complete control over their subjects.

The revival of the castle, especially the restoration of castle culture is closely linked to the rise of Romanticism. Following the Industrial Revolution, upper-class liberal idealists became more frustrated with the centralized despotism of the French King, leading to the Revolution. British, society being dominated by the bourgeoisie, still has the new generation of the hierarchical society choking it, the so-called high society being both avaricious and pretentious. With Austria and Germany still using the feudal system, Europe is desperate for a drastic makeover.

Romanticism echoes many Enlightenment ideas; they both embrace freedom and equality, but the romantic is adamantly against the excessive rationalism of Enlightenment thought. He, instead, promotes naturalism, individuality, imagination and emotions, resisting the conventional. Reminiscence of the troubadours' fervor and the medieval knights' passion and valor, and praise the pastoral beauty of folk art lead to a transformation of the castle image.

The castle was originally a symbol of the dark, bloody, barbaric feudal system of the Dark Ages. However, the castle has become a mark of myths and folklore. Even more so, castles became an architectural model for new buildings, this so-called period of the "Gothic revival" is not only the revival of the urban Gothic, but also of the rural Gothic.

To this day, Enlightenment and Romantic thought still dominate the mainstream values of European society, so, castles are beautiful not only because of construction, but also because the relevant culture has made them special. Today, key castles are World Cultural Heritage sites or protected by the government, and are also tourist sites. Some castles have been used conservatively and made into museums, art galleries, research institutes, and hotels. Many popular tourism spots in Europe are where castles are concentrated, such as, Provence, Tuscany and the Rhine Valley, and is by no means coincidental.

浪漫主义思潮对启蒙运动思潮既有呼应，也有反弹，它们都推崇自由平等，但浪漫主义反对启蒙思想中过分的理性主义，更为推崇自然主义、个性化，在文化风格上强调个人情感和想象力，反感"正统"的化身古典主义，怀念中世纪骑士那种感性化的英雄情怀、行吟诗人的激情澎湃，赞美田园美和民间艺术，这些取向导致古堡形象的彻底转变。

古堡原来是封建制度的象征，是"黑暗的中世纪"中阴森、血腥、野蛮的标志。而这时，古堡却成为民族神话、民间童话的标志；古堡更作为一种建筑风格被新建筑广泛采用，这时期所谓的"哥特复兴"不仅复兴城市哥特式，也复兴乡村哥特式。

至今，启蒙和浪漫主义思想仍然是欧洲社会的主流价值观，所以，古堡不仅是因为自身的造型美，还因为相关文化而一直受到人们特殊的喜爱。今天，重要的古堡都是世界文化遗产或国家保护文物，也是旅游对象；部分古堡被保护性利用，成为博物馆、美术馆、研究所等，有的成为古堡酒店，欧洲几个最受欢迎的旅游度假区都是古堡集中区，如法国普罗旺斯、意大利托斯卡纳、德国莱茵河谷等，这绝非是偶然的。

在建筑学方面，古堡风格继续作为一种设计风格流行，不仅是外形，内部装饰和环境更流行，其作为自然主义、浪漫主义的符号，也为所有设计、文化创作领域提供灵感。城堡和要塞作为一个建筑分类而存在，对应的有专门的学术研究机构，定期召开国际学术会议。

法国拉罗谢尔的城堡和要塞　　The castle and fortress of La Rochelle, France.

德国摩泽尔河边的古堡
The castle along the Moselle River, Germany.

瑞士阿尔古堡
Al castle, Switzerland.

意大利波皮古堡，在此举办的东西方之间的城堡要塞国际会议上，意大利教授在介绍中国山西皇城相府的情况（Francesco Maglioccola 摄影）
Popi castle, Italy. Italian professor is introducing the Chinese castle –Huangcheng Xiangfu in Shanxi at an International conference on Eastern-Western fortress (Photo by Francesco Maglioccola).

4. 古堡建筑的要素和美感

古堡在人们心目中有一种约定俗成的形象，它们外墙高大，外窗较少，女墙带雉堞，有凸出的塔楼等。

更具体的，古堡外围一般还有壕沟或护城河，古堡的外墙有的是独立的，厚度要使顶部能供防守人员活动；有的与藏兵洞结合；有的就是房屋，只是其外墙格外厚重。由于在防守时侧面攻击非常重要，城墙多有外凸的马面、碉楼，至少在转角处都有外凸设施。外凸设施有的是在外墙上空出挑，并高出墙体。许多古堡的围护墙体不止一层、一环，多层、多环墙体的组合形式更多。

有些古堡相当于一座集中式建筑，更多的古堡有内部院落，院落中可能有独立的高大塔楼，这种塔楼是最终的防御地，在外围失守后还可以继续坚持，等待援救，所以，那里通常是古堡主人的住所。塔楼可以有很多座，有的塔楼与外围建在一起。所谓集中式古堡往往就是一座独立的大塔楼，或称碉楼。

内院内可能还有其他房舍，用于仓库、武士住所等，古堡里通常还有牢房，有时牢房在塔楼顶，更多的是地牢形式，无怪乎，许多古堡后来都变成了监狱。

在有古堡的贵族们越来越富裕，身份感越来越强后，他们越来越重视古堡的美感和威武感，用以彰显自己的权势、地位、财富、品味。在古代，这样做也许是必要的，但在浪漫主义运动之后，人们会认为连残垣断壁都会美，而且

西班牙塞哥维亚城堡
Segovia Castle, Spain.

法国鲁昂城堡残存的"圣女贞德塔"
The Joan of Arc Tower, the remaining part of the Rouen Castle, France.

In the study of architecture, the castle style continued to be popular as a design style, not only in the profile, but the interior decoration and environment are even more popular as the symbol for naturalism and romanticism, providing inspiration to every aspect of design. The castle and fortress classification exists in Europe, resulting in academic and research institutions and regular international conventions.

4. Elements and Beauty of Ancient Castles

Most people have a conventional image of a castle in their minds: a tall external wall, smaller windows, battlements on parapets, protruding towers, and so on.

More specifically, on the castle periphery, there usually lies a trench or moat and curtain walls that are stand-alone, the tops of which are wide enough for soldiers to stand on. Some are used for soldiers to hide in while others simply add an additional obstacle, albeit a very thick one. Turrets, flanking towers and corner towers are important in maximizing the number of people available to defend the area. Some of these towers are overhangs protruding out of the curtain wall. Many castles not only have one curtain wall with towers stemming from them, that have many layers of such walls decorated with defensive towers. Some castles are consolidated structures while most others have courtyards, some of which lead to a stand-alone tower that is fortified and defensible as a last refuge. The tower is also usually the domicile of the lord of the castle. Many towers can exist within the castle walls, but more consolidated castles have such towers incorporated into the outer wall. Therefore, such consolidated castles are often just one big stand-alone tower, or Diaolou.

Within the courtyard there may be other premises used as warehouses or soldiers' quarters. Castles often also have prison cells in tower and dungeon form, so it is no surprise that many became prisons.

法国卡昂的诺曼底公爵城堡
Duke of Normandy castle, Caen, France.

法国昂热城堡
Angers Castle, Angers, France.

有特殊的美，大片实体墙面不仅不会单调，其质感、沧桑感都有美感。至于精致的古堡，就更不用说了。

上述古堡建筑的要素构成决定，古堡必然有更起伏的轮廓线、变化更丰富的造型、更自然多样的姿态，加之古堡多建于山地、坡地，结合地形的作品更加变幻莫测，在风景的衬托下一定更加动人。

在古堡的防御功能越来越淡出之后，原来的壕沟或护城河会成为一种环境艺术，内院也会变成花园，中世纪后期和文艺复兴之后的古堡本来就多附加有花园、农庄，这样，古堡最容易成为欧洲式的世外桃源。

不仅在视觉上，在精神层面的审美需求上，古堡逐渐消除了负面因素，唤起正面效应。在人们可以更心平气和地看待历史以后，在作"文明的野蛮人"的共识下，古堡原来的封建性变成了独立性，原来的粗野历史变成了有趣的故事，与文化名人、历史事件有关的古堡还有特殊价值。

在中国，古城的城墙曾经不是文物，而是落后的象征，需要拆除。长城倒是最重要的文物，还是民族、国家的象征，但也有人说过长城是封闭性的象征，在辩证法流行的年代，很少有人用辩证法指出，军事堡垒既是防御设施也是进攻设施这一军事常识。古堡无人重视，直到有位著名作家用宁夏银川市附近的两座古代古堡式军营遗址作影视拍摄地后，类似古堡的美感才开始引起公众注意。

山西皇城相府，中国北方最大的私家城堡
Huangcheng Xiangfu, Shanxi – the largest private castle in northern China.

Castle nobles, feeling increasingly affluent and titled, became aware of the beauty and might of castles that demonstrate their own power, position, wealth, and taste.

In ancient times, such behavior may have been requisite to prosper, but after the Romantic movement, one would think even the ruins of large areas of solid wall that not only still stand, but have texture and history that are especially intriguing and beautiful, not to mention finer, whole castles.

With the aforementioned architectural elements, an outline of a castle consists of changing, undulating shapes that are natural and diverse in nature. This, coupled with the various backdrops in which castles are built, from mountains to slopes on a hillside, makes castles unpredictably striking.

As castles lost their defensive function, trenches and moats became environmental art and inner courtyards transformed into gardens. The additional gardens and orchards that sprang up after the late Middle Ages and the Renaissance effected with ease the transition of castles into an out-of-this-world getaway.

Not only in appearance, but on the spiritual level, castles' negative associations gradually came into a positive light. When one could look on the history calmly, people now see the once feudal castle as a symbol of independence, the violent history as an interesting anecdote. The rich culture and history behind castles also make them that much more valuable.

In China, ancient city walls were not considered cultural artifacts but symbols of a mindset of the past and needed to be demolished. In reality, the Great Wall is a universal symbol of this nation, and while some have said that it symbolizes enclosure, others have suggested that it is both a defensive and offensive structure.

No one paid attention to castles, and it wasn't until a film was shot in two ruins of castle barracks near Yinchuan in Ningxia Hui Autonomous Region that people came to see them as beautiful.

二、中国古堡建筑的特点

1. 特殊社会情况的产物

秦代以后的古代中国与罗马帝国有许多相似之处，政府主要通过城市系统管治社会，城市里有官僚和军队的住所，有时军队有独立的住所，但多数军营往往与城市是相似的格局。由于城市的格局有礼制等方面的严格规定，所以一般都比较呆板和千篇一律，即千城一面。

皇帝住在帝国最大的城市中，与官僚和军队将领在城市中的住所是基本不设防的府第不同，皇帝一般住在城中城里，虽然这种皇城比较小，但和多数军营一样，并没有城堡的特征。而如北京紫禁城之大，就更是城，不是堡了。

有时在战争地区，朝廷会设置坞壁、寨堡一类的建筑，这种建筑只强调军事性，不用理会城市格局，根据自然情况兴建，所以这类建筑会有些古堡意象。但也因为它们只强调军事性，使它们普遍过于简单化，也很难完整保存下来。

山西平遥古城的城墙，古代城市的城墙一般在夯土外有包砖
The city wall of Pingyao County, Shanxi. Ancient city wall has external brick layers wrapped around the rammed earth.

军事戍堡建筑标准较低，城墙一般只有夯土
Army garrison fort, normally built by rammed earth.

宁夏现在的"西部影视城",过去是明清军队的戍堡
Present, Western movie set city, Ningxia; Past, Ming and Qing army garrison fort.

青海草原上游牧民族修的古城
Castle in the grassland of Qinghai built by Nomads.

Section 2　Distinguishing Traits of Chinese Ancient Castle Architecture

1. The Products of Unique Societal Contexts

There are many similarities between ancient China after the Qin Dynasty, and the Roman Empire. The government governs through an urban system with the city officials and military personnel residing in the city, though sometimes soldiers' barracks are separate, and has similar layout as city. There exists a rigid ritual pattern that is maintained when being constructed that all cities look exactly the same.

古代中国一直有名义上的贵族，上等贵族还有名义上的封地，但他们与欧洲中世纪的贵族不同，他们对自己的封地基本上没有政治、军事管辖权，也没有在自己的封地里建造城堡的权利。他们多作为特殊形式的官僚住在封地里的城市中。而欧洲城市多数在中世纪时就是自治城市，贵族只好住自己建的郊野城堡。

同时，中国历朝历代也均有大贵族、大官僚、大氏族、大地主为彰显自己的势力、保卫自己的财产建造民间性的坞壁、寨堡、土围子，但他们多是中央政府的打击对象。只有在乱世中，中央政府自顾不暇，这时他会鼓励民间自保，还指望民间力量勤王，民间的城堡建筑就会大量出现。

在汉代至唐代各王朝末期的乱世中，史料记载那时的中国遍地都是坞壁建筑，数量当不会比欧洲的城堡少。坞壁应该是中国人当时对古堡建筑的称谓，从唐代末年至清代末年，类似建筑的称谓改成了寨、堡，所以，后来才发展出了"古堡"一词。

由于年代久远和朝廷的不宽容，中国明代晚期之前的民间古堡几乎荡然无存。明末天下大乱时，民间一部分人起义了，还有一部分人聚众建寨堡自保，就如《明史·流贼列传》中有记载"所过，民皆保邬堡不下。"是说李自成的起义军所过之处，当地老百姓担心被抢被杀，都躲在"邬堡"中不出来，李自成对此无可奈何，他最终就是在查看一座民间寨堡时遭寨堡中的人袭击而死。

明末，欧洲的火炮由葡萄牙人传到中国，明朝政府军配备后重创过农民军和后金军。后金军（不久更名清军）很快也使用火炮，在火炮面前，民间寨堡的防御性就不是政府军担心的事了。应该主要是这个原因，虽然清朝政府也实行集权专制制度，但对民间存在寨堡比较宽容，没有都当成所谓土围子拆除。

浙江临海古城，主要用于抵抗倭寇
Seaside castle in Zhejiang, mainly used for defending Japanese pirates.

The emperor, usually resided in the largest city of the empire. Although the domiciles of said general and bureaucratics are basically undefended, the emperor usually lives in a smaller, defended city within a city with no particular features of a castle. Even the Forbidden City in Beijing, while large, is more city than castle.

Sometimes in war zones, the court will order castles and fortresses to be put up, emphasizing military efficiency rather than conforming to the standardized urban pattern, thus naturally resembling the castle image. Unfortunately, because the design is overly simplistic and militaristic, these fortresses are difficult to fully preserve.

Nobility in ancient China are only nobles in name. While upper nobility are granted land in which they live and act as bureaucrats, but they were really different from the Medieval nobles of European, they have no political or military jurisdiction within the fief land and therefore cannot construct castles within their fiefdom. The majority of European cities during the Middle Ages are autonomous cities, in which nobles can build their own castles.

Meanwhile, the Chinese dynasties have great nobles, bureaucracies, clans, and landlords that demonstrate their power and defend their property by constructing castles, fortresses, and fortified villages, and operate against the central government. Only in a troubled world, when the central government hasn't a leg to stand on, will it encourage private self-protection and expect civillians' show of strength, from which sprouts ample construction of castles by loyal citizens.

During the troubled times at the end of every dynasty from Han to Tang, the records show construction of Wubi no fewer than the number of castles in Europe. Castles were called "Wubi" in the early time, from the Tang Dynasty to the Qing Dynasty, with similar constructions termed Zhai (stockades), Bao (forts), and later, castles.

Due to the long time and intolerance of the court, vernacular castles essentially did not exist before the late Ming Dynasty. During the chaos at the end of the Ming Dynasty, some held

重庆钓鱼城的城门，南宋军队曾经在这里用炮火将蒙古帝国的蒙古大汗击伤致死，从而影响到欧洲、中东的历史形势

The city gate of Diaoyu City in Chongqing. The army in South Dynasty used cannon to insure then kill the Khan of Mongolian Kingdom, thus affecting the historic events in Europe and the Middle East.

湖北山区的寨堡遗迹　广东东江边的古堡
The relics of castle in the moun-　The castle near the Dongjiang River in Guangdong Province.
tain region in Hubei Province.

到太平天国和捻军起义时,政府又鼓励民间建堡自保。中国现存的民间古堡多是明末、清末建造的,一些天高皇帝远的地方,政府管治力度不足,许多大家族为了安全,在清代中期所谓盛世中,也能建城堡。一些城堡的建造目的也有彰显门第的意图,也有建筑形式习惯使然。

2. 现存中国古堡的分布和主要类型

中国现存古堡最多、最集中的地区从北至南分别为古代农牧分界线沿线,山西省汾河、沁河流域,闽粤赣3省中国历史中中原南迁人群与南方人群杂居地区。

另外,西南地区长时间实行土司制,这种制度有些封建性质,土司多有自己的寨堡。然而在清代中期实行"改土归流"政策之后,湘、黔、川、滇等地的多数土司寨堡不复存在,改为政府的屯堡,一些屯堡后来变成民间的住所,得以保存下来的同时,也增添了古屯堡的古堡意象。

大西南更深处的羌族、藏族居住区域封建制度一直延续到20世纪中期,相应地,那里有更多的古堡保存下来,只是大型古堡多为官方所有。

农牧分界线沿线古堡多数是宋代以后的军屯、戍堡,它们多是长城防御体系的一部分,在长期战争后,多只剩下黄土夯筑的残垣断壁。一些古堡可能原来就是军民混居的,也可能是在被军队放弃后,有民众住进去,总之,还有人居住的古堡就仿佛还活着。这种古堡中古堡元素的体现主要是城墙、城楼。

多数古堡的现存建筑历史一般不会早于明代,在更早时期,中国还流行过一种"台"式建筑,虽然没有建筑实物留下来,但可以想象,这种建筑立于高台之上,有相当强的防御性,与许多建在陡峭高地上的古堡在形式上是类似的;

uprisings while others cooperated to build their own castles for protection, as documented in *History of the Ming, A Brigand's Biography*, it tells of Li Zicheng's rebels passing through towns where their citizens feared robbery and murder, all hiding in castles refusing to leave. Li Zicheng was helpless against these people, and he eventually was killed in a fortress he was scouting out.

At the end of Ming, the artillery of Europe, provided by the Portuguese, reached China. After the Ming army acquired cannons, it sought out and fought against peasant armies and the army of late Jin. The late Jin army (soon renamed the Qing army) started to use cannons soon after. Vernacular castles and fortresses no longer remained threats to the government. Although the Qing government also implemented a centralized autocracy, it tolerated the existence of vernacular castles and not to demolish them, because their defensive capabilities could not withstand cannon fire. In response to the uprising of the Taiping and Nien, the government again encouraged self-protection through castle building. Most of the castles in present day China were built in the late Ming and Qing dynasties. Some larger families that lived outside the reach of the government were able to build their own castle during the middle of the Qing Dynasty, sometimes called the Golden Age. Some castles were constructed to highlight the gate, while others rose from architectural habit and personal style.

2. Main Types and Distribution of Existing Chinese Castles

China has three regions where castles are numerous and highly concentrated. They are, from north to south: the line dividing pastoral land from farmland in ancient times, the Shanxi basin where the the Fen and Qin Rivers flow, and the Fujian, Guangdong and Jiangxi provinces where in history northerners immigrated south to these provinces.

In addition, the southwest region operated under the chieftain system, the feudal nature of which caused need of Castles. In the mid of Qing Dynasty, however, due to the bureaucratization of native officers in Hunan, Guizhou, Sichuan, Yunnan and other places, tribal castles were replaced by the government's military fortress. Some, later, became private residences, which, in addition to preserving, added to the imagery of the ancient castle.

Further southwest in Qiang and Tibetan regions, the feudal system endured until the mid-20th century, effecting the preservation of more castles.

The majority of castles along the pastoral dividing line are garrison forts from after the Song Dynasty. They were part of the Great Wall's defense system, and after the long war, only ruins of dirt walls remain. Some of the castle may be the original is the soldiers and civilians mixed, may also give up after the military, people live in, In short, the castles only seem to endure if people still live there. The main castle elements of these castles are the city walls and towers.

如果是位于宫殿中的台，就与城堡中的塔楼形式类似。因为筑高台工程量巨大，所以只有朝廷和大贵族官僚可以筑台。"台"式建筑到底是什么样子，除了古画上的形象可参考，黄河宁夏段沿线现存几座高庙建筑，其中应该有"台"式建筑的记忆。此外，日本古堡，即日本古代"城"中的"阁"应该也是一种记忆。

山西省汾河、沁河流域古堡与闽粤赣3省客家、闽人、潮人、粤人的古堡是中国现存古堡的主体，也是本书重点记述的内容。

西藏定日的老城民居和山顶城堡式的寺庙
Ancient residential houses and hilltop castle-style temple in Dingri city, Tibet.

俯瞰广武城
The bird's eye view of Guangwu city.

西藏萨迦的设防民居和城堡式的寺庙
Walled village and castle-style temple in Sajia city, Tibet.

The existing architectural history of these castles generally do not predate the Ming Dynasty, before which there was a wave of terraced architecture. Although none remain standing today, one can imagine this type of building standing above the high-terrace and have fairly strong defensive ability, similar to castles built on steep terrain. Terrace structures within palaces are similary to castle towers. Because of the large workload to build such terrace, so only members of the court and bureaucratic nobility can live in terraced buildings. What terraced architecture was actually like, building in the end what it was like, in addition to the old paintings on the image reference, the segment of the Yellow River in Ningxia houses several high temples that retain some memory of terraced architecture. In addition, Japanese castles, or pavillions in ancient Japanese city, also resemble the terraced buildings.

The castles in the Fenhe and Qinhe River basin in Shanxi, and in the three provinces of Fujian, Guangdong and Jiangxi of the Hakka, Fujian, Chaozhou, and Guangdong people comprise the main body of the existing castles in China, and are also the focus of this book.

宁夏平罗高庙
Tall temple in Pingluo, Ningxia.

宁夏中卫高庙鸟瞰图
Tall temple in Zhongwei, Ningxia.

山西高台式的设防堡院
Terraced walled village, Shanxi.

中国现存古堡分布图
Map of China's existing castles.

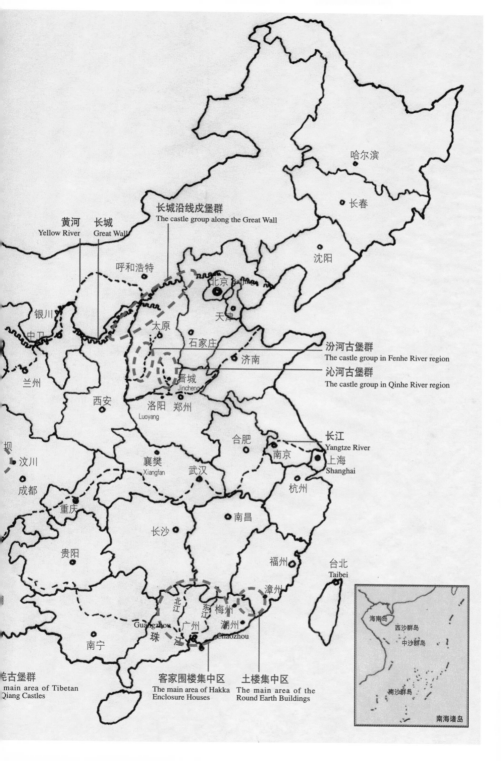

3. 与欧洲古堡建筑的异同

中国的古堡建筑形式基本上是自我独立发展的，在清末之前，与欧洲的交流不多，而在魏晋南北朝之前，交流应该更少。

但是，中国古堡与欧洲古堡之间却有一些惊人的相似之处，如广州市出土的东汉明器表现的"四角楼"式古堡在客家古堡中有大量实物存在，是中国古堡的一种基本形式，而这种四角楼在欧洲也是古堡的一种基础形式。很难想象、也无法证明，这种形式是谁传给谁的。

这样的现象只能说明：人类各族群之间一方面存在着各种不同，一方面也存在太多的共性，在各自独立的实践中，都会提炼出古堡建筑的"原型"。各自也都会发现：城墙顶部需要雉堞，城墙上需要提供侧面保护的马面，角楼和塔楼对防御至关重要等技术细节。在社会环境有共同点时，也会出现相似的景象，如制度上有封建性、地广人稀的古代西藏地区，就曾经出现欧洲中世纪式的人居景象：山顶城堡与城镇聚落互为依托。

中欧古堡最大的不同主要有3点：第一，欧洲古堡建筑的主体一般就是古堡建筑自身，设计是整体性的，而中国古堡的建筑主体多是古堡墙体内的院落；第二，来自各民族、各时代建筑风格的不同；第三，建筑材料的不同，欧洲几乎不用夯土技术。

葡萄牙里斯本的贝伦塔，主塔就是四角楼形式
Belém Tower, Lisbon, Portugal, the main tower is a four-corner-tower form.

意大利乡间的四角楼古堡式府邸建筑
Four-corner-tower castle-house in the countryside, Italy.

3. The Similarities and Differences From European Castle Architecture

Chinese castles developed independently from those of Europe. There was scarce interaction with Europe by the late Qing Dynasty, and even less before that during Weijin Northern and Southern Dynasty.

However, there are some striking similarities between Chinese castles and European castles. For example, in Guangzhou City, an excavation revealed the funerary objects of Eastern Han Dynasty, which is a model of Sijiaolou castle, that is similar to many existing Hakka castles, which is also a fundamental castle design in China, as well as Europe. It's hard to imagine or prove who had the idea first.

That this prototype can appear in both parts of the world shows how people of any ethnicity are fundamentally the same while being vastly different in their creativity and independence. Independently, each can discover that city walls need battlements, walls need, turrets, and towers at the corner or flank are all essential to defense of the castle. When the social environments share similarities, the scenery will also familiar, such as in the institutionally feudal and sparsely populated ancient area of Tibet, where there once stood a medieval European scene: the symbiotic hilltop castle and surrounding town settlements.

The difference between Chinese and European castles is in three main ways: first, the main body of the European castle building is usually the castle itself, while that of Chinese castles are often the courtyard within the castle walls; second, differences in architectural style of each nation and era; third, differences in building materials, in Europe, densely packed earth is rarely used if at all.

广东龙川比较标准化的四角楼
The standard four-corner-tower building in Longchuan, Guangdong.

4. 民间古堡的精彩和文化价值

虽然中国的古堡目前没有欧洲古堡有名,但它们同样精彩,特别是民间古堡,它们都是大家族的家,像欧洲贵族的古堡一样,古堡要表现主人的形象,所以在文化艺术和美学上不惜投入,这必然使民间古堡的品质远远超过纯军事戍堡。

深圳高碉楼在水中的倒影　　The reflection of tall tower (Diaolou) in water, Shenzhen.

民居建筑在中国古建筑中有特殊价值，在不受过多约束的情况下，民居会包含古人更多的创造性；在中国正史往往是帝王家事史的情况下，民居可以记录下许多正史缺失的信息。

民间古堡是最特殊的一种民居，作为一种在集权专制制度下不合法，只能在特殊情况下存在的建筑形式，它在各方面肯定会更加不同寻常。

汉代的班固认为，孔子说过"礼失求诸野"这样的话，真是圣人之言。在历朝历代末期，不仅礼失要求诸野，治失也要求诸野。民众为了生存，各寻生路，古堡是藏身之处，建堡反映的也许只是求生的本能，但毕竟是保护生命和文化的主动措施，也是民间自治意识萌发的一种现象。

4. The Marvel and Cultural Value of Vernacular Castles

Although Chinese castles are currently not as well-known as European castles, but they are still wonderful, especially vernacular castles. They are all large-family homes, like the castles of the European aristocracy. Because the castle needs to represent its master, it often heavily invested in art and aesthetics, which is bound to place the quality of vernacular castle far above purely military garrisons.

Vernacular architecture is of particular value in ancient Chinese architecture; without too many constraints, rural residential areas better preserve the ancient people's creativity. Since Chinese official history is often the history of the imperial family, these families can record much of history that does not coincide with what the imperial family deems relevant or true.

Vernacular castles are unique dwellings, since not legal under the centralized autocracy, these architectural forms exist only in exceptional circumstances, which makes these castles all the more unusual. Ban Gu of the Han Dynasty believed that Confucius's words: "Retrieving Lost Rites from Barbarians" are words of a sage. At the end of every dynasty, not only the lost rites but also the rule of law are only retrieved from the indigenous people. Because the people wish to survive, the castles that were built may reflect such, but also became an initiative for the preservation of culture and life, which gave rise to a sense of autonomy.

5. 历史沿革概要

中国古堡的前身坞壁至少在秦朝末年的乱世中就出现了，这从成语"作壁上观"出自《史记·项羽本纪》可以佐证。在记录东汉末年乱世的史籍中，坞壁更大量地出现，如《三国志》载："筑郿坞，高与长安城埒，积谷为三十年储，云：'事成，雄踞天下；不成，守此足以毕老'"。说的是董卓修了一座名为"郿坞"的坞壁，以期进退自如。另外，成语"坚壁清野"也是出自三国时期。

结构更复杂的碉楼式东汉明器
High tower (Diaolou) styled burial objects with complex designs.

山西平遥现存的高碉楼
An existing Diaolou, Pingyao, Shanxi.

河南陕县出土的东汉明器，碉楼上有弓箭手，碉楼下有骑兵
The burial objects of Eastern Han Dynasty unearthed in Shanxian County, Henan. Archers in the Diaolou, cavalries at the foot of the Diaolou.

河北汉墓壁画中的高碉楼形象
High towers (Diaolou) image on murals in Han tomb, Hebei.

5. Chronological Summary of History

The Wubi, a predecessor of the castle, appeared in the chaos of the end of the Qin Dynasty, as can be corroberated by the set phrase said: sitting on the Wubi and seeing, which was in *"Historical Records: Xiang Yu Basic Annals."* Historical records show the appearance of many Wubi during the chaotic ends of both the Eastern Han Dynasty and the Sui Dynasty. The armies fought on the fields and hiding in the Wubi, and the general people all went into the castles for hiding, the enemy couldn't grab any useful goods and materials.

湖北鄂城出土的三国四角楼式明器
Four-corner-tower burial objects in the Three Nations era, Echeng city, Hubei.

广州出土的东汉四角楼式明器
Four-corner-tower burial objects in Eastern Han Dynasty, Guangzhou.

甘肃武威出土的东汉明器，形式为一座有中间望楼的四角楼中竖立起一座碉楼
Burial objects in Eastern Han Dynasty, Wuwei, Gansu. A high tower (Diaolou) built in the centre of a four-corner-tower with watchtower.

隋末天下再次大乱,《旧唐书·屈突通传》载:"通令显和夜袭文静,诘朝大战,显和纵兵破二壁,唯文静一壁独完,然数入壁,短兵接,文静中流矢,军垂败,显和以士疲,乃传餐食,文静因得分兵实二壁。会游军数百骑自南山还,击其背,三壁兵大呼,奋而出,显和遂溃,尽得其众。"《新唐书·王世充传》载:"时百姓皆入壁,野无所掠。"

那么坞壁是什么样子的呢?没有当年的建筑实物遗存。查《故训汇纂》,坞字有"小障也、壁垒也、里也、营居曰坞"等意,应为封闭、有防卫、有些军事化居住意味的寨堡类建筑;壁可能指另外一种防卫性建筑,它应该比坞更小,军事性更强,而"壁"字也可能是用来表明坞有高墙厚障防护。结合其他史料的描述,我们可以认为坞壁应比一般的城池小,在建制上应比城池等级低,一般也独立于城池之外,为小一些的、等级低一些的小城或堡寨。

除了文字记载,各类出土文物中也有坞壁形象的记录,特别是从汉代至三国的墓葬中出土的陶制明器和壁画上都有一些坞壁的形象,这些形象在中国现存的古堡上都有所体现。如广州市郊出土的有4个角楼、前后望楼的汉代明器,即前述的"四角楼";湖北省鄂城市出土的有带角楼、望楼高墙围护的大院落的三国明器;高大塔楼形象出现在河北省安平县逯家庄汉墓壁画上,还有甘肃省武威市、河南省陕县出土的东汉陶楼。这些文物上的形象应该是对贵族、官僚私家坞壁的描绘,军事和纯民间坞壁应该比之简单、简陋一些。

唐代以后,坞、壁的名称逐渐向堡、寨转变,名称虽然变了,所指的建筑形式应该变化不大,现存的明清时期的堡、寨,形式仍然反映着出土文物上汉代坞壁的形式。只是能保留到今天的高楼,都不是木结构的,至少是土木结构的。

广东始兴县满堂围
Mantangwei, Shixing County, Guangdong.

Well, what does Wubi look like? With no physical remains of such structures, research in The Corpus of Ancient Glosses explains that "坞" (wu) is a type of castle architecture that encloses and acts as a garrison. "壁" (bi) could mean another form of defensive construct, smaller and more battle-hardy than "wu". The word "bi" may also be used to decribe a "wu" that had a thick wall of protection. In combination with other historical descriptions, we can deduce that "Wubi" should be smaller than the average city and has a lower construction grade. While generally independent of cities, they are smaller, lower grade towns and stockade castles.

In addition to the records, various types of archaeological finds in resemble the Wubi, especially ceramic funerary objects unearthed in the tombs from the Han Dynasty and the Three Kingdoms and murals that depict images of the Wubi. These images also exist on Chinese castles still standing today. The funeral objects of Han Dynasty,unearthed in the suburbs of Guangzhou are model of Sijiaolou which have four corner towers and two watching towers .Unearthed in Echeng City, Hubei Province were funeral objects of Three Kingdoms with turrets, watchtower, high-walls surrounding a large courtyard. The images of tall Keep towers appear in the Han tomb murals in the Lujiazhuang, Anping County, Hebei Province. In Wuwei City, Gansu Province, and Shan County, Henan Province, the Eastern Han ceramic House was excavated. These relics depict the aristocracy, bureaucracy, private Wubi; military and native Wubi are often more humble and smaller in size.

After the Tang Dynasty, the name of the Wu, Bi gradually shifted to "castle." Although the name of such structures changed, the architectural form remained more or less the same. The existing castles of Ming and Qing dynasties closely resembles the archaeological finds from the Han Dynasty. Towers still standing today are at most half brick, half wood; no purely wooden towers have endured through the ages.

第二章 黄河流域的民间古堡
Chapter II　Vernacular Castles of the Yellow River Basin

山西皇城相府河山楼
The Heshan Tower in Huangcheng Xiangfu, Shanxi

一、历史文化背景

1. 农牧交替、胡汉混居

西部半农半牧人群不断东迁变成以农业经济为主的趋势，将黄河流域造就成为中国最早的经济文化政治中心。紧接着，北方的半农半牧人群不断南迁，挤压中原人进一步南迁，使中国的经济文化中心不断南移。这样的趋势一直持续到清代。同时，农耕和游牧人群对以黄河、长城为主轴的农牧分界线地区的争夺始终在进行，造成该区域常年战争不断，农耕人群除了以长城为防御主体，也兴建大量堡垒。在北宋时，为了抵御北方契丹人和西北方党项羌人，黄河流域的农耕人基本都住在城堡里，南宋时，城堡防线南移至长江流域。

这样的历史情况似乎与农耕人多筑城，游牧人多筑堡的规律矛盾，然而，首先是上述规律未必全面；其次，历朝历代与游牧人群对抗的农耕人群往往是刚从游牧人群转变的；最后，如半农半牧的契丹人、党项羌人等也有大量城堡。总体上，游牧人群是进攻者，而其进攻的根本动力其实是他们的生活不如农耕人稳定，一旦他们的进攻得手，占领了农耕人的土地，他们多数就会转变为农耕人，转身抵御新出现的游牧人，如此一来，黄河流域早已是胡汉混居区。

山西省博物馆内展出的《山西汉末西晋时期少数民族分布图》，可见当时的情况
The distribution of national minorities from late Han to Western Jin periods in Shanxi province, Shanxi Museum.

农耕人自称汉人，自古有文化优越感，制造"华夷之辨"、"农牧之辨"等，称游牧人为胡人。而且，在"一丘之貉"这个成语产生前，孟子就用"貉国"来指代游牧国家，针对有人提议中国中土的农耕国家应向周边的游牧国家学习，减少统治系统、苛捐杂税、繁文缛节等，孟子不以为然，他认为像貉国这种没文化的"行国"向国民征收二十分之一的税是够了，但中土国家不行，必须取十分之一的税才够政府维持国家文明水平。

博物馆内胡人形象的陶器　The pottery of Barbarian image in the museum.

山西皇城相府　Huangcheng Xiangfu, Shanxi.

Section 1　The Historical and Cultural Background

1. Agricultural/pastoral region, Hu and Han mixed living

Western agricultural and pastoral populations constantly moved eastward into the trend of mainly agricultural economy, created the Yellow River Basin of China's earliest economic, cultural and political center. Then, the agricultural and pastoral people of the north continue shift south, squeezing the Central Plains people further down south, so that China's economic and cultural center constantly moved southward. This trend continued into the Qing Dynasty. In the meanwhile, the main agro-pastoral farming and nomadic groups compete on the dividing line areas along the Yellow River and the Great Wall, always in progress, causing the region's perennial war. The farming population not only used the Great Wall as a defensive mainstay, but also constructed of a large number of fortresses. In Northern Song Dynasty, in order to resist the northern Khitan and the north-west of the Xiang Qiang group, the farming population of the Yellow River Basin basically lives in the castle. In the following Southern Song Dynasty, castle defense line moved southward to the Yangtze River Basin.

The above historical fact seems inconsistent with the rules of thumb that farming group prefers building fortification, and nomadic people tend to build Forts. However, first of all the above rules may not be comprehensive; Second, the farming population of the dynasties against the nomads were often settled former nomads themselves. Again, farming and animal husbandry population, such as the Khitan people, and the Qiang people also have a large number of castles. After all, the fundamental driving force of the nomads attack is in fact their lives is not as stable as the farming people. Once the nomadic attackers succeeded in their offense and occupied farm land, most of them will transform into farming people, and turned against the new influx of nomadic people. Therefore, the Yellow River Basin has long been a mixed settlement of Hu and Han.

Farming people claiming to be Han Chinese, the ancient cultural superiority, manufacturing "Hua Yi," "Pastoral Distinction" considered the nomadic people as barbarians. Moreover, before the creation of the idiom "*Small beast comes from a hill* ", Mencius used "Raccoon country" to refer to nomadic countries. He dismissed the proposal of central China farming country should learn from the nomadic country, i.e. simplifying the ruling system, reducing exorbitant taxes and levies, and cutting red tapes. Mencius thinks the national tax collection of twentieth of income is enough from the no-culture "traveller" communities for the raccoon country, but not for the central China nation. The central China must have taxes as the tenth of income which is necessary for government to maintain the national level of civilization.

2. 改朝换代、绝境求生

事实上，农耕包括前游牧人建立的半农半牧的王朝政府最终都会把税收提高到三分之一、二分之一。另外，"行国"之所以没文化，农耕人认为是它无"宫室宗庙祭祀之礼"，城堡当然算不得宫室宗庙，而建像样的宫室宗庙就要加人民的税，所以农耕王朝经常会因为税负过重使百姓没饭吃而发生危机，最终王朝覆灭。

每当危机深重时，黎民百姓必遭涂炭，豪门大户也难瓦全，想活下去的人，唯一办法就是武装自己，民间武装一般没有能力建城，只有建城堡自保。

现在，除了黄河、长城防线一带，黄河流域一些地区的山顶上也有古堡废墟，它们中甚至有唐宋遗迹，总之都是历史中的危机期，特别是改朝换代时期有人群为逃生进入深山，在地势险要处建的城堡遗迹。

二、山西的特殊性

中国北方现存的古堡多集中在山西省，而且山西的古堡品质最高，这是有一些必然性的。

农牧文化、农商文化特殊的关系，特殊的风土，特殊的历史际遇，缔造了山西古堡的辉煌。

1. 最丰富的半农半牧区

山西省的地理形式非常特别，对于黄河下游地区来讲，山西地处高地，在黄河随意泛滥的年代，山西相对安全，古代东迁的人群往往会在山西驻留。炎帝部落就是东迁部落，而炎帝是中国从游牧过渡到农耕的标志，他的姓——"姜"仍带着游牧人的印记，但他也是中国的"神农"。山西东南部的古上党地区俯瞰黄河下游，史称这里是神农活动的主要地区。

山西西南部同样是中国最古老的一片文明之土，所谓"尧都平阳、舜都蒲阪、禹都安邑"都在这一地区。随后，周朝最重要的诸侯国之一晋国的中心也在这里，后来才北移至山西中部。如此深厚的农耕文化积淀使山西在受游牧文化冲击时不会丧失文化根基，反而能使游牧文化的活力有效注入。

山西的西北边界基本与农牧分界线重合，虽然有高山大河等天然屏障，有长城，但挡不住各游牧人群进入山西，山西的自然条件也适合半农半牧，所以

2. Regime change, desperation to survive

In fact, farming, including farming and animal husbandry, established by the former nomad dynasty government will eventually shift the tax to one-third and up to the half. The farming nation reasons the "Traveller nation" has no culture, due to it has no palace, temple and ritual worship tradition. In order to build decent palaces and temples, a lot of money is needed from increased people's taxes. Farming dynasty is often collapsed because of the crisis of no food for people to survive under the heavy tax burden, leading to the final destruction of the dynasty.

Whenever the crisis deepens, the common people shall die by being trampled on, the only way for the wealthy rich to survive is to arm themselves. Civil armed force generally does not have the ability to build a city, only to build a castle to protect themselves.

Till now, castle ruins can be found along the defense line of the Yellow River, and the Great Wall, also in areas of the Yellow River Basin hilltops, some even dated back to the Tang and Song dynasties. Generally speaking, castle ruins are remains from the crisis period in all history, especially regime change period, built in the difficult terrain area by the fleeing crowd escaped into the mountains.

Section 2 Particularity of Shanxi Province

The existing castles in northern China are more concentrated in Shanxi Province, and Shanxi castles have the highest quality, inevitably. Special relationship between the culture of agriculture and animal husbandry, agro-business culture, special endemic, and special historical fate, created Shanxi Castle's brilliant.

1.The most plentiful farming/pastoral areas

Shanxi Province, has a very special geographical form, located in the highlands free from persistent flood caused by the Yellow River. Compared to the lower reaches of the Yellow River, Shanxi is relatively safe, therefore, ancient people moving eastwards tend to settle down in Shanxi. The Yandi tribal is one of the east-moving tribes. Yandi is the symbol of transition from nomadic to farming group. His surname - "Jiang" is still with the mark of the nomadic people (his name in Chinese character is composed of two parts, one is meaning sheep, another is meaning women), but he is also China's "Shennong" – the farming god. The Gushangdang area of the southeast of Shanxi overlooking the lower reaches of the Yellow River, is known as the main areas of Shennong activities.

The southwest of Shanxi is also part of China's most ancient civilization, the so-called "Yao's capital is Pingyang, Shun's capital is Puban and Yu's capital is Anyi" are all in

游牧人进入山西后往往乐于在此定居或常来常往。秦汉时期，山西北部是农牧交锋的前线，从汉末至隋代，山西就基本上是匈奴、羌胡、鲜卑等游牧人的乐园了。李氏唐朝起兵于山西,而隋唐两朝的统治者本身的游牧人因素是非常重的。唐末至清末，只有北宋和明朝时期，山西属汉人王朝统治。也就是说，汉代之后，山西在一半以上的岁月里是游牧人、半农半牧人的天下。

虽然不太平，但总体上山西的地理还是非常封闭的，相比黄河中下游其他地区，在动乱年代，山西所受破坏相对较轻，中国元代前的古建筑在山西存量最大的现象就是这么形成的。山西深处还有许多局部封闭的小天地，即使山西也有战乱，这些小天地要么不易被发现，被发现了也吸引不来大队人马，建城堡才有防御意义，以城堡的一般规模是防御不了千军万马的。

2. 晋商

建城堡是需要钱的，特别是私人建城堡，或需要参与的人多，或需要富豪人家。古代的富豪人家，要么是大官僚，要么就是大地主、大商人，山西

山西古建筑的特性使其在与其他风格的砖石建筑结合时更易于融合。左2图分别为新绛县和阳曲县的教堂建筑
The charactor of ancient architecture in Shanxi make it easy to compromise with the other style brick architecure. The two pictures are churches in Xinjiang county and Yangqu county.

the region. Subsequently, the center of Jin, one of the most important vassal states in the Zhou Dynasty is also situated here, and later on moved to the central Shanxi in the north. With such a strong farming cultural heritage, Shanxi did not lose its cultural roots under the influence of nomadic culture shock, but to make the vitality of the nomadic culture effectively injected instead.

The northwest border of Shanxi mostly coincides with the agricultural/pastoral dividing line, although there are mountains and rivers and other natural barriers, as well as the Great Wall, but can not stop the nomads entering into Shanxi. Shanxi natural conditions for agricultural and pastoral, nomadic people into the Shanxi are often willing to settle or exchange frequent visits. Qin and Han dynasties, northern Shanxi is the front line of pastoral encounters, from the late Han Dynasty to the Sui Dynasty, Shanxi is basically a paradise for the nomadic people of the Huns, Qiang Hu, Xianbei minority groups. Lee Tang Dynasty, raised an army in Shanxi, the rulers of the Sui and Tang dynasties still heavily influenced by their nomadic ancestry/inheritance. From the end of the Tang Dynasty to the end of the Qing Dynasty, Shanxi was ruled only by Han's emperor during the Northern Song Dynasty and the Ming Dynasty period, In other words, since the Han Dynasty, Shanxi was governed by nomadic people, agricultural/pastoral for more than half of the historic period.

Although it was not absolute in peace, but in general Shanxi is geographically enclosed, compared to the middle and lower reaches of the Yellow River region. In the troubled years, Shanxi had suffered relatively light damage, so that the ancient buildings stock dated back and beyond the Yuan Dynasty are still exist in Shanxi. There were many the partial closure region in the depths of Shanxi, even if there were wars in Shanxi, these regions can not easily be found, nor attracted to large army deployment. Only building a castle in these enclosed regions makes sense in defense term, because a normal scaled castle can not defend huge mighty military force.

2. Shanxi Merchants

Building a castle is in need of money, especially private funded project, either needs plenty of participating contributors, or the solid financial support from wealthy rich. Ancient wealthy rich, are either the high-ranking officials, or the big landlords, big merchants. Shanxi was a small region, no big landlords, the profits of agriculture was also limited, but there was no lack of big merchants. After-rich merchant family still passed on farming and cultivation culture as the tradition. On the one hand, pursuing the acquisition of land, on the one other hand, allowing future generations to study, passing the imperial bureaucracy exams, and eventually get bureaucratic richer. Shanxi's best castles were mostly built by these rich families.

地少人多，没什么大地主，农业的利润也有限，但山西不乏大商人，致富后的商人家庭多依农耕文化中耕读传家的传统，一方面购置土地，另一方面让后代读书，通过参加科举产生官僚，产生官僚后就会更富。山西最好的古堡多是这种人家建的。

农耕人是严重抑商的，游牧人则往往反之。晋人多从商，有为生活所迫的，恐怕也有传统使然的。晋人从商还有资源优势，在中国古代，盐铁两业是国民经济的重中之重，通常由政府垄断，但政府总是能把最好的行业搞赔钱，没办法的时候就让民间商人来收拾残局，民间商人往往能妙手回春。山西有盐池、有煤矿铁矿，同时，山西处在有财富但商品匮乏的蒙古草原和物产丰富的中原、江南之间，生产、贸易都可以搞，后来又率先搞起金融业，因此，山西自古出富商大贾。

山西砥洎城，古代的藏兵洞变成现代的羊圈，也是一种农牧交融
Ancient places of hiding soldiers become modern sheep fold, a blend of agriculture and animal husbandry, Diji city, Shanxi.

山西皇城相府古堡复杂的城墙构造
The complex structure of castle walls, Huangcheng Xiangfu, Shanxi.

Farming people were seriously suppressing commerce trading, nomadic people were often the contrary/opposite. Most Shanxi people are merchants in business, either forced by the pressure of living, or dictated by the tradition. Jin dynasty business took advantages in rich resources. In ancient China, two industries, i.e. salt and iron, are the top priority of the national economy, usually controlled by a government monopoly, but the government always can manage the best of the industry to lose money. Then, the government let private traders to clean up the mess, private traders often can maneuver the trade returning profit with ease. There are salt ponds, coal and iron mines in Shanxi, it also located between rich but commodity-poor grasslands of Mongolia and property wealth of the Central Plains, south of the Yangtze River. It was easily to engage in production, trade, and later on, Shanxi also took the lead into financial sector, therefore, Shanxi nurtured rich merchants since ancient times.

上2图：山西祁县曹家大院，前部院墙较矮，但有雉堞，后部高大如城墙
Two above: The Cao family compound, Qixian county, Shanxi.
Low front part of the external wall with battlement, tall rear part of the external wall like a city wall.

皇城相府展出的表现传统冶铁工艺的模型
The model of the traditional iron smelting process in the exhibition, Huangcheng Xiangfu.

3. 农牧商多元文化下的山西古建筑、古堡

山西古建筑存量大、质量好，首先得益于半农半牧区气候相对干燥，土木建筑易于保留；其次，山西多煤，便于烧砖，砖木建筑比土木建筑坚固；再次，山西多富商，质量高的建筑必多，何况，山西人爱盖房子是公认的；最后，山西建筑品质高、美，主要是得益于多元文化之间既矛盾又相容。山西古建筑既有农耕文化的细腻保守，也有游牧文化的自然奔放，还有商业文化的华丽灵活。

游牧人还在游牧的时候，他们一般不从事建筑，而他们一旦定居，建筑的热情并不比农耕人低，而且在天苍苍野茫茫的环境中养成的性格还会使他们在建筑时表现出比农耕人更高的豪情和更大的胆量。同时，游牧人承认农耕文化整体上更先进，他们可以虚心学习农耕人的技术，所以游牧人往往能建出更伟大的建筑，山西几座最惊人的古建筑多是在游牧人的王朝时期建的。

至于农耕人爱筑城，游牧人爱建堡的现象，主要原因应该是农耕地区人口众多，更适合建城市，游牧地区人口少，适合建较小的城堡。山西北部游牧区的属性更重，所以多城堡、少城市。北部现存的古堡多为军事戍堡，仅右玉县，

沁河流域的大阳古城，城门与设防堡院的碉楼结合
Dayang ancient city, Qinhe River basin. The city gate joint together with Diaolou in the fortified courtyard.

3. The multicultural excellency: ancient architecture and castles in Shanxi

Large stock of good quality ancient architecture exist in Shanxi, because firstly, climate is relatively dry in the farming/ pastoral areas, vernacular construction is easy to retain; Second, Shanxi has coal for manufacture bricks easily. Brick and wood construction/ structure is sturdy than the clay construction/structure; Third, Shanxi has many wealthy rich as well as many high-quality construction, it is well known that Shanxi merchants love to build houses; Finally, Shanxi architecture processes high quality and beauty, mainly thanks to the multi-cultural contradiction and compatibility. Shanxi ancient buildings embedded both delicate conservative farming culture, and natural and unrestrained nomadic culture, plus the gorgeous flexibility of the business culture.

Nomads generally do not engage in construction while in their transient travel living, and once they settled, the enthusiasm for building is not lower than the farming people. With the character formed in the wild amongst the vast green environment, they often show more innovative and adventurous in the construction than farming people. Nomads recognize the farming culture as a whole is more advanced, they can humbly learn farming technology, they often able to build more grand buildings. Consequently, several amazing ancient buildings in Shanxi were built in the dynasties governed by the nomadic people.

山西沁河流域的周村古镇，城门和城墙的构造
Ancient town in Zhoucun village, Qinhe River basin, Shanxi. The city gate and city wall.

这种古堡就存有 100 多座，它们和野长城一样，有一种特殊的美，但在建筑学方面意义稍逊。山西最好的古堡集中在晋中和晋东南，那里农耕区的属性更重，城市较多，古堡几乎都是商人建的私人或小团体的住所。

4. 活化石现象

山西现存古建筑不仅年代古远，还保留下更古远的建筑构造模式，如明清时代的建筑中会有汉唐的构造和形制。产生这一现象的原因可能是山西不仅建筑保留得好，建筑传统保留得也好；此外，地理上的封闭性多少都会导致一些文化上的封闭性。

具体到古堡，汉唐坞壁的形式要素和宋代以后已不普遍实行的里坊制度都被山西古堡保留下来。至今，山西的许多地名还在显示这里曾经是坞壁密布的地方，也显示山西人不是随波逐流不断改地名的人。现在的山西还有许多地名是××坞、××壁、××里、××坊，这显然是坞壁、里坊流行时代留下的。

山西灵石红门堡内部的里坊格局
Li-fang layout in the Hongmen castle, Lingshi county, Shanxi.

The phenomenon of farming people love to build city, nomadic people love to build castle, mainly due to populous farming areas are more suitable to build the city, a small population of nomadic areas, suitable to build a smaller castle. In northern Shanxi nomadic character is more dominant, so there are many castles, less small cities. In the north amongst existing castles mostly military garrison fort were found. Merely in Youyu County, there are over 100 of these types of castles. Similar to the wild Great Wall, there is a special kind of beauty, but termed as low achievers in the architectural significance. The best castles of Shanxi concentrate in the regions middle and southeast of Shanxi, where farming areas are more dominant with more built cities, the castles are almost all built as private or small group residence by businessmen.

4. A living fossil phenomenon

Shanxi existing old buildings, not only dated far back to the ancient era, but also retain more ancient mode of building construction, for example, the buildings of the Ming and Qing dynasties, contain the Han and Tang dynasties structure and shape. The reason for this phenomenon might be that the building retains good architectural characters together with construction traditions. Moreover, the geographical closeness would lead to certain extend of the cultural closeness.

The form of Wubi in Han and Tang dynasties, together with the Li-fang unit system which no longer widely practiced after the Song Dynasty, is well preserved in Shanxi castles. Nowadays, many place names in Shanxi are still kept as the original names, and shows that Shanxi did not follow the general trend to constantly change place names. The Shanxi many place names are the Wu, Bi, Li, Fang, which are obviously inherited from the popular Wubi and Li-Fang era.

上2图：沁河流域郭壁内的堡院均以"里"、"坊"命名
Two above: castle courtyard within Guobi, Qinhe River basin.

三、沁河流域民间古堡

沁河为黄河支流，发源于山西中部的太岳山，一路南流，在河南省汇入黄河。沁河穿越的山西沁水、阳城二县煤炭资源丰富，当地自古有冶铁业，加之地处晋豫交通要道上，便于贸易，至少在明末的商业繁荣期（有历史学者认为是中国古代的资本主义的萌芽期），因此产生了大量富商。在随后的动乱期，因为陕西饥民常常来劫掠，明朝政府自顾不暇，为了自身生命财产安全，这些富商纷纷建起私家城堡，或联合乡里将村落设防，由此产生的一批古堡部分地保留下来。

1. 沁水县窦庄

沁河古堡最密集的地区是沁水县端氏镇至阳城县北留镇一段，端氏镇这个名字还留有封建制度遗痕，指周代端氏封地。战国末期，这一带是秦赵长平之战时秦军的驻地，据说周边的武安、屯城、王离、马邑等地名都是因此留下来的。这些地方当时都是秦军的堡垒，或许，他们为这片土地留下了古堡的因子。

今天的端氏镇以煤炭工业兴旺，在古代，这里还有缫丝业，富甲一方。窦庄在端氏镇以南，沁河西岸，因可能是东汉征服匈奴的名将窦宪之后的北宋将军窦璘家族由陕西迁此定居而得名。至明代，张氏家族成为窦庄大族，张五典、张铨父子均为万历年进士，其中张铨后来在辽东为抵御后金殉国，是袁崇焕的前辈。

张铨殉国时，张五典已去官还乡，儿子的死使他意识到大明将亡，天下将乱，已近古稀之年的他遂带领家人和乡人将窦庄建成城堡，据说张五典在这个小村庄复制了京城北京的模式，城堡分内外城，外道城墙环村，内道城墙环张家主宅，故窦庄有"小北京"之称，张家主宅仿佛紫禁城。城堡于明崇祯二年（1629年）建成，距今已有380多年历史。

皇城相府的侧面城墙
Side view of the city wall, Huangcheng Xiangfu.

Section 3 Vernacular Castle in Qinhe River Basin

Qinhe River, one of the Yellow River tributaries, originated from Mountain Taiyue in central Shanxi Province, and flows south all the way, to join the Yellow River in Henan Province. Qinhe River runs across Qinshui and Yangcheng counties rich in coal resources, local metal fabrication industry since ancient times, coupled with its location on major traffic routes joining Shanxi and Henan provinces, in ease of trade during the commercial prosperity in the late Ming period (history scholars believe that period was the ancient Chinese capitalism infancy), resulting in a large number of wealthy businessmen. In the subsequent period of unrest, the Ming government was hardly able to deal with the often plunder by Shaanxi famine refugees. In order to protect their own lives and property, the wealthy had been built private castles, or joint township vulgar to the village castle, retained the resulting number of castles.

1. Douzhuang village in Qinshui County
The most densely populated areas of the Qinhe River castles in Qinshui County are from Duanshi town to Beiliu town in Yangcheng County. The name of Duanshi, also left/retained the traces of the feudal system, referring to the fief of Duan family in Zhou Dynasty. At the end of Warring States Period, this belt was the military camping sites for the battle between Qin's and Zhao's armies in Changping region. It is said that, Wu'an, Tuncheng, Wangli,

窦庄堡的前门
The front gate of Douzhuang castle.

堡内里坊式的巷道，端头有上城墙的楼梯
Li-fang-style roadways inside the castle, Douzhuang castle.

窦庄堡的背面城墙　Rear side of the Douzhuang castle.

历史证明张五典颇有先见之明，他死后不久，陕西饥民就杀到沁水，各村镇皆遭杀掠，而窦庄在张铨夫人的主持下，乡民据堡固守，毫发无损。有成功经验参考，加上也是明军将领的张铨之子张道浚鼓励民间建堡，以减轻明军的防御负担，于是这一地区在明末建起了几十座民间城堡。而窦庄堡因此有"沁河第一堡"之称。

今日的窦庄因年久失修略显破败，外城墙基本不存，只剩少数城门。内城墙还比较完整，近10米高的城墙里是一个小里坊格局的院落群，主院中有一座3层高的楼房，类似高楼在沁河古堡中都有，它们为这种设防村落或院落式的城堡增添了至关重要的古堡要素。

2. 沁水县郭壁

由窦庄沿沁河向南，不到1公里即到紧邻河堤的郭壁，如果结合河堤形成临河的城墙，郭壁不仅易守难攻，还可控制沁河河谷。而郭壁的地名就说明这里自古是一座城堡。如今，郭壁的城墙也已不存，但村内尚存一座堡中之堡和数座堡院。

Mayi and other names were thus remained since then. Fortifications of Qin Dynasty were placed all over these places, they perhaps/likely formed the determinants/factors of the castle in this piece of land.

Today, the coal industry is thriving in Duanshi town, it was also wealthy with silk reeling industry in ancient times. Douzhuang town is located in the south of Duanshi town, at the west bank of Qinhe River. It is named after the Northern Song Dynasty General DouLin family (whose ancestor was the General Dou Xian defeated Huns in the Eastern Han Dynasty) moved from Shaanxi and settled there. Till the Ming Dynasty, the Zhang's family becomes the influential family in Douzhuan village. Zhang Wudian, and his son Zhang Quan both were the Scholars in Wanli period. Zhang Quan, the predecessor of Yuan Chonghuan, was martyred in the battle against the Later Jin in the east of Liaoning province.

When Zhang Quan martyred, Zhang Wudian had resigned from official duty and returned to his hometown, his son's death made him aware of the Ming Dynasty was approaching perish, and the world will be chaotic. Almost seventy years of age, he then led his family and local people to convert Douzhuang village into a castle. It is said that Zhang Wudian replicated the layout of capital Beijing in this small village. The castle consists of concentric inner and outer walled rings, the outer walls wrapped around the village, while the inner channel walls ring around Zhang's main house, just like the configuration of the Forbidden City. The castle was built in the second year in Chongzhen era (1629) in Ming Dynasty, dating back over 380 years of history.

History has proved that Zhang Wudian was quite prescient, soon after his death, Shaanxi hungry immediately stormed Qinshui County, towns and villages were invaded and slaughtered. Under the auspices of Mrs. Zhang Quan, the villagers of Douzhuang village, fought by the castle fortress and whole village remained intact. With previous successful experience, also encouraged by the Ming army generals Zhang Daojun, son of Zhang Quan, for reducing the defense burden of the Ming army, dozens of private castles were built in this area in the late Ming Dynasty. Douzhuang Castle was therefore named as "the first Castle in Qinhe River region".

Today's Douzhuang slightly run-down due to disrepair, the outer wall basically does not exist, leaving only a small number of city gates. The inner walls is relatively complete, nearly 10-meter-high walled courtyards formed by small Li-fang system layout. There is a three-storey building in the main courtyard, which is common to have similar high-rise structure in the Qinhe River castle. It is also adds the crucial elements for this kind of walled village or courtyard-style castle.

从城墙的破损部分可见内部夯土，门闩为石制
Internal rammed earth, visible from the walls of the damaged part of the city wall, the gate latch made from stone.

暗道是沁河古堡内常设的一种防御设施，窦庄的旗杆院与内堡之间就有暗道。著名作家赵树理曾经在这座院落居住、写作
Secret passageway is one of the usual defensive measures in Qinhe River castles. There is a secret passageway between the flag pole courtyard and the inner courtyard in Douzhuang. The famous writer Zhao Shuli used to live and write in this courtyard.

窦庄内的其他院落
Other courtyards in Douzhuang.

The underground passage is a norm defense facility in Qinhe River castle, there were underground passages connecting the Douzhuang village's flagpole homes and inner Fort. Famous writer Zhao Shuli once lived in this courtyard and writing.

2. Guobi village in Qinshui County

Guobi village located southwards along Qinhe River, is less than 1 km from Douzhuang village and close to the riverbank. Its outer wall formed in combination of embankment made Guobi not only easy to defend off invading attacks, but also had control of the Qinhe Valley. The name of this place-Guobi, proved it was an ancient castle. Nowadays, the outer walls of Guobi does not exist, but a central castle and several castle courtyards in the village still exist.

Guobi was built in greater scale than Douzhuang, also most likely richer than Douzhuang, so there is an ancient saying "Gold Guobi, Silver Douzhuang". Long time ago, Guobi was a trading town, there was 2 km-long commercial street along the river, the town-houses facing the street are shops, inner house has layout of Li-fang. A gate of Li-fang called "Sanhuaili" located between the shop verandahs, inner streets behind this gate connected

郭壁堡中之堡"青缃里"全景
Overview of the castle-in-castle, Qingxiangli, Guobi Castle.

郭壁的规模比窦庄大，好像从前也比窦庄富，故古有"金郭壁，银窦庄"之说。从前郭壁是一座商镇，临河有2公里长的商业街，镇上的房子临街是铺面，里面则是里坊格局。名为"三槐里"的一座里坊门夹在有外廊的店铺之间，坊门里由内街连接各个院落，这些方形院落虽然不是古堡，但外墙高大，外窗很少，角部有所谓"插花楼"（也称看家楼）高起，不仅有一定的防御性，还与所谓"四角楼"相似，所以可将之视为一种堡院。

　　内街尽头的小山坡上，是堡中之堡"青细里"，平日，这里是书院，有战乱时，可以作为内堡。古堡外墙高大，但内部并不觉得因围墙过高而压抑，这是因为堡内地平面比外部高五六米，而城墙直接从堡外地平面升起，下半部是挡土墙，上半部才是围墙，而城墙的外表是一体的。古堡大门内有陡峭的台阶通向堡内地平面，一片台地上有两座堡院。

郭壁名为"三槐里"的里坊门内侧
Inside view of the Li-fang gate named Sanhuaili, Guobi castle.

街头高地上的"青细里"古堡
Qingxiangli castle on the high ground in the street.

to each courtyard. These square courtyards are not castles, but with the tall external wall, a little external windows, plus the so-called "Chahualou"(also watch tower) in the corner. It served defensive function to certain extent, similar to the so-called "Sijiaolou", so it can be seen as a castle court.

A castle within castle named "Qingxiangli" was built on the hills at the inner end of the street, in normal days, it is used as a school, during wars, it can be a fort. Viewing from the inner courtyard, tall castle external walls do not bring feeling of too high fenced nor suppressing, this is because inner floor level is 5-6 m higher than the external castle field level where external walls rising directly from. The lower half of the external wall is retaining wall, only the top half is the actual surrounding wall, overall the external wall appears as a whole tall wall. Steep steps lead to the inner castle platform within the castle gate, a platform has two castle courts.

里坊内的堡院
A courtyard in the Li-fang village.

古堡入口内的台阶
The steps at the entrance of the castle.

古堡内的院落
A courtyard inside the castle.

院落内的碉楼式正房
The main house (Diaolou) in the courtyard.

3. 阳城县砥洎城

沁河流入阳城县境后不久，在润城镇外形成一片湖面，湖水中远远可见一座古堡，它叫砥洎城，因为沁河古称洎水，古堡建于河心一块如中流砥柱般的巨石上而得名。

砥洎城名为城，其规模确实在城镇和城堡之间，形式也介乎于古城和古堡之间。

中国古代社会的官制只到县一级，所以县城是官方最小的城，都设城墙。县城以下，除了军事寨堡，官方就不负责筑城了，有战乱时，民间或自保，或躲进县城、府城之类。民间自己修城堡时，大户人家可自行其是，村落、小商镇也比较容易协调，大的城镇就不好协调了，没有权威在，谁出钱？谁出力？谁来管理？所以古代设防的镇最少，如郭壁在镇一级的设防上就比较潦草。润城镇的主体也没有设防，但在镇主体的西北侧，有一座设防的砥洎城也相当于一个小镇的规模，而且城墙高大，至今保留比较完整，非常难能可贵。砥洎城能兴建，估计还是得益于润城镇有超强的经济实力，沁河的冶铁业以润城最为兴旺。

当年陕西饥民来袭时，润城一带遭到极大破坏，连明万历年进士、刑部侍郎张慎言的儿子都死于非命。为了不使悲剧重演，润城的大富商们协商起来显然比大明朝朝廷上的朋党间容易得多。砥洎城具体的建造年代不得而知，依古堡中现存的明崇祯十一年（1638年）"山城一览"碑看，应该在此之前大体建成。

砥洎城的水门　　The water gate of Diji city.

坩埚城墙局部　　Part of the crucible city walls.

3. Diji town in Yangcheng County

Shortly after flowing into the border of Yangcheng County, Qinhe River forms a lake outside Runcheng town. From far distance, a castle can be seen standing in the lake. It is so called Diji town, which was originated from Qinhe's ancient name - Ji Shui, the castle named because of being built on a huge boulder like the mainstay in the river.

Diji town to be called a town, its size is indeed between a town and a castle, its form is also in between an old town and an ancient castle.

Bureaucracy system of ancient Chinese society only extended down to the county level, so a county is the official smallest city, with allocated defensive city walls. Below the county level, the official is not responsible for building the fortification, except for the Military Fortress. If there is war, civil population either self-protection or hide in the county or city. When civil folks building their own castle, rich families could go their own way, small towns and villages were also relatively easy to reconcile, but the larger towns had poor co-ordination. If there is no authority in control, Who pays? Who contributes? Who will manage? So that rare ancient fortified town exist, e.g. a fortification of Guobi in the town is more primitive. The main body of the Runcheng town set no fortification, but the northwest side of the main town, a walled Diji town is also equivalent to the size of a small town, and the tall walls, has retained a relatively complete, very commendable condition. The successful construction of Diji town had relied on the powerful economic strength, Qinhe River region metal fabrication industry was most prosperous in Runcheng town.

临水的双层城墙　Water front double-layer city walls.

古堡整体平面形状呈不规则椭圆形，原来三面临水，正门在临陆地一面，朝润城镇，这一面的城墙大量使用冶铁业报废的坩埚砌成，独特而坚固；水中城墙上还开有一座水门，门内有瓮城，城墙上密布藏兵洞，瓮城的地平面比城内主体地平面低十余米，这样，水城一面就有了两道城墙。

两道城墙上面高高的、有插花楼的院落是清代数学家张敦仁的故居。成片民居的布局也是里坊制的，共有10个坊，除了住宅，坊间还有关帝庙、文昌阁，朝水的一面有黑龙庙。

与砥泪城相邻的屯城、刘善两村，在同时期也建有城堡，只是今天只存遗迹。

瓮城内部
The inner view of barbicans city.

窑洞式的藏兵洞
Cave-style hiding place for soldiers.

▶ 城内防御性很强的街巷
The city streets with strong defensive functions.

坩埚城墙下的院落
A courtyard under the crucible city wall.

In the year of Shaanxi hungry strikes, the area around Runcheng was greatly damaged, also a scholar in the Wanli era in Ming Dynasty, the son of powerful official, was killed. In order to prevent a repeat of the tragedy, Runcheng wealthy gathered in consultation together, which was clearly much easier than assembling the Great Ming court cronies and officials. The Diji town's construction period is not known, and pursuant to the existing monument of "mountain town at a glance" dated in Chongzhen 10th year (1638), the construction of the castle should have generally completed before that time.

Overall, the castle has an Irregular oval shape, three walls facing the water, the main entrance located in the land side, towards the Runcheng town. This section of the walls were made by a large number of scrapped smelting iron crucibles, forming a unique and sturdy structure; a water gate in the walls opened out onto the river, barbicans placed at the inner side of the gate, with densely nested hiding place for soldier in the walls. The ground level of the barbicans was more than ten meters lower than the city main ground platform, so that there were two walls in the Watertown side.

The courtyard high above the two walls also containing the Chahualou was the former residence of the Qing Dynasty mathematician Zhang Dunren. The common residential layout was also Li-fang, there were 10 Fang in total. Apart from homes, there were Guandi temple, Wenchang Pavilion, and the Black Dragon Temple placed toward the water side.

Near Diji town, castles were also built in the two neighbouring villages, Tuncheng and Liushan, in the same period, but only the remains still there.

刻有砥泊城平面图的石碑　Engraved plan of Diji town on a monument stone.

4. 阳城县郭峪

沁河的一条小支流樊溪在润城镇汇入沁河，溯樊溪向东数里，可见唐代古刹海会寺，寺中两座古塔，造型都非常独特，特别是高的一座，建于明代嘉靖年间，其砖砌的八边形下部完全是古堡碉楼的样子。

继续溯樊溪，在溪谷较宽阔处，樊溪西岸有一座古城规模的郭峪古堡，东岸山坡上，还有一座小古堡——侍郎寨。如果站在郭峪古堡背后的山上继续向樊溪上游望，溪谷里还有更壮观的"皇城相府"，还有数座古庙点缀在山间，这一景象不仅是沁河古堡群最精彩的画面，也是中国古代田园最精彩的画面之一。

郭峪是个大村，但由于没有设防，当陕西饥民于崇祯五年（1632年）来袭时，根本抵挡不住，村中死伤上千人。这之后，在村中的老进士、大商人的倡议下，村人有钱出钱、有力出力，3年后，环村的城墙就建了起来，稍后，由王姓富商按"皇城相府"河山楼的模式又建起了一座塔楼，这种塔楼与汉代明器、壁画上表现的坞壁高塔非常相似，与欧洲中世纪古堡中的塔楼也非常相似，作为村中最后的避难所。

从郭峪望皇城相府，近景是郭峪的汤帝庙和豫楼
The perspective view of Huangcheng Xiangfu from Guoyu castle. Tang emperor temple and Yu Tower are in the close range view.

4. Guoyu village in Yangcheng County

Fanxi Creek, a small tributary of Qinhe River, joins Qinhe River at Runcheng town ship. Travelling eastwards few miles along Fanxi, the Haihui Temple of Tang Dynasty would appear in the view. The two pagodas in the temple both have very unique shapes, especially the tall one, built in Jiajing era in the Ming Dynasty. Its lower part of octagonal masonry duplicates the design of castle towers.

Continue travelling along Fanxi reaching the wider section in the valley, you will see an ancient city scale castle, Guoyu castle, at the west bank, on the hillside of the east bank, there is also a small castle—Shilangzhai. If you stand on hill behind Guoyu castle and continue to look towards the upper reaches of the Fanxi Creek, there are even more spectacular "Huancheng Xiangfu" in the valley, as well as several temples dotted in the mountains. This scenery is not only the most exciting image of the ancient Qinhe River castle group, but also one of the most wonderful pictures of Chinese ancient pastoral.

Guoyu was a large village, but in the absence of fortification, it could not withstand the struck of the Shaanxi hungry in the Chongzhen 5th year (1632), over thousands of villagers were killed and injured. After that attack, under the initiatives set by old scholars and big merchants

郭峪的城墙外侧
The external view of the city wall, Guoyu.

城墙内的街道
Street within the city walls.

城墙内侧
Inner side of the city wall.

汤帝庙与一座城门相结合
Tang emperor temple joins together with a city gate.

豫楼正面　The front view of Yu Tower.

海会寺有碉楼状塔座的琉璃塔
Haihui Temple has colored glaze tower with the base of defense tower (Diaolou).

汤帝庙与城墙的结合
The joint section of Tang emperor temple and the city wall.

汤帝庙侧后方的城门，带一座小瓮城
The city gate behind the side of Tang emperor temple, it includes a small barbicans city.

正月十五时郭峪张挂的彩灯
The display of lanterns during traditional Lantern Festival's in Guoyu.

琉璃曾经是阳城县的优质产品，豫楼下原来有一些古代真品，现已被盗
Coloured glaze ceramic was once a high-quality products in Yangcheng County. Some genuine ancient coloured glaze found in Yu Tower, but has been stolen.

从侍郎寨望郭峪主体
Outlook of the main body of Guoyu from Shilangzhai.

塔楼名为"豫楼","豫"字可通"预"字,有预防之意,也可暗指河南省,这一带人的祖辈许多是从河南迁来的。楼内有水井、暗道,避难时可以生存,也可以进退。

从构造上看,郭峪更像是一座设防村落,而不是人们惯常认为的古堡,但因为有豫楼,还有与一座城门结合的汤帝庙,使它的局部非常有古堡的意趣。我们觉得欧洲的古堡好看,很大程度上是因为在欧洲古堡的体积构成中有高塔和体积比一般民房大得多的教堂或会堂,有它们才易于产生好构图。郭峪的高塔是豫楼,教堂便是汤帝庙。汤帝庙本身像是座大古堡中的小古堡,高大门楼左右有更高大的、碉楼状的钟鼓楼,特别是它与城门楼、瓮城、城墙的结合,使这一群体呈现出层层高墙中塔楼林立的造型,完全是古堡的结构。

5. 阳城县"皇城相府"

在崇祯五年(1632年)的那场劫难中,幸免于难的除了窦村人,还有原来郭峪的大户之一陈氏家族。

陈氏家族以煤炭、冶铁业致富,家族中也有人在明朝为官。随着家势壮大,原来在郭峪村中的住所日益紧迫,陈家就在郭峪斜对面的山坡上另辟新宅。可能是因为得到了关中大乱的消息,领会到窦庄修城堡的必要性,陈家在新居旁修建了一座高塔楼,取"河山为囿"之意命名为"河山楼"。崇祯五年时,楼的

上2图:皇城相府内部
Two above: inside views of Huangcheng Xiangfu.

in the village, the villagers raised money and gathered defense efforts, three years later, the ring walls of the village were built up. Later on, a wealthy rich surnamed Wang, also built a tower mimicking the Heshan Tower in "Huangcheng Xiangfu". This tower is very similar to the Wubi tower on the Han Dynasty funerary objects and murals paintings, is very much like the tower of the medieval castle in Europe as well. It functioned as the last refuge in the village.

The tower called the "Yu Tower", "Yu" means "prevention", also hints "Henan Province". Many of the ancestors of the people in this region were originated from Henan. Yu Tower is equipped by wells, underground passages; people can survive in this temporary refuge, also can advance and retreat.

From the structural point of view, Guoyu is like a walled village, not the usual castle people recognized. However, the Yu Tower, plus a Tang Temple combined with a city gate, makes it partially a very charming castle. We admire the beautiful castles in Europe, largely because the composition of the volume of European castles. Tall towers together with church or city hall (which volume is much larger than normal houses) are just easy to make a good composition of architecture. Guoyu's tower is the Yu Tower, the church is the Tang Temple. Tang Temple itself likes a small castle within the castle block, nearby the tall gate there is the Bell and Drum Tower like bigger tower. In particular, the combination of this tower with gatehouses, barbicans and protective walls, forming the scheme of towers lined with layers of walls, just like the multi-layered structure of an ancient castle.

5. Huangcheng Xiangfu in Yangcheng County

Villagers of Douzhuang, as well as one of the original Guoyu large Chen family survived during the catastrophe in the 5th year of Chongzhen's reign.

The Chen family got rich from coal and metal fabrication industry, a member of the family was also an official in the Ming Dynasty. As Chen household constantly growing, the original residences of Guoyu village urgently need expansion. Chen family then starts to build new houses on the hillside diagonally across Guoyu. Overheard the rumor of chaos in Guanzhong region, also understood the necessity of building the castle in Douzhuang, Chen built a high tower next to the new house. The tower named as Heshan Tower taking the meaning of "Rivers and Mountains Surrounded Park". In Chongzhen 5th year, the main body of the tower had just built, merely in time to defend the inflowing strikes by the Shaanxi hungry. Heshan Tower had saved the life of Chen family and many nearby villagers. On that occasion, even the larger fortified cities in the east, Dayang town and Zezhou City, had been defeated and broken. It was quite remarkable that Douzhuang and Chen family survived and overcame the attacks.

古堡边的陈家私家花园
Chen family's private garden beside the castle.

从南侧俯瞰皇城相府
The perspective view of Huangcheng Xiangfu from the south.

窑洞式藏兵洞前的民间音乐表演
Folk music performance in front of the cave-style places for hidden soldiers.

主体刚刚建好,就赶上陕西饥民来袭,河山楼救了陈家和附近不少乡民的命。那一次,连东面的设防大城大阳镇和泽州城都被攻破,窦庄和陈家能保全,实在难能可贵。

这之后,陈家又用一圈近乎长方形的高大城墙将新居和河山楼围在里面,形成一座大城堡。清代康熙年间,陈家子弟陈廷敬官至文渊阁大学士,位极人臣。为体现家族气象,陈家又在明代城堡前面加建了一圈包容着两座功名牌坊和官式院落的城墙,从而形成一座蔚为壮观的双城古堡。明代部分名为"斗筑居",清代部分名为"中道庄","皇城相府"是近期开发旅游时所用的名称。

无论是从建筑的历史年代、记录的文化信息,还是古堡的建筑品质来看,皇城相府都是中国现存古堡中的佼佼者。

第一,它最经典地体现了古堡建筑的特征。

皇城相府建在山地上,内部各种建筑分布在几层经过人工修整的台地上,城墙则依山势逶迤而上,这样,从整体看,一层层的水平结构被城墙的斜线收拢,再与河山楼的垂直线交织,不仅制造出美妙的轮廓线,还展现出无尽的细节,气势磅礴下有田园的怡然,浑然一体中有生活的细腻。河山楼及城墙上的众多角楼集中体现了坞壁建筑的特征,特别是河山楼,其通高28米,外形方正,墙体实际上是下部厚,上部薄,这样,楼越下部越坚固,越上部内部空间越大,既美观又实用,楼内同样有水井、暗道。后部城墙结合台地形成3层窑洞形藏兵洞,窑洞的圆弧、实体性,与空透性的木结构建筑之间形成的对比,也大大提升了造型

从陈氏宗祠回望"斗筑居"城门
View of the Douzhuju city gate from Chen family's ancestral temple.

古堡内的监禁室
A detention cell in the castle.

After that, Chen family built another ring of tall walls in rectangular shape wrapping around the new homes and Heshan Tower to form a large castle. In Kangxi era, the offspring of Chen family, Chen Tingjing, ranked as the official WenYuan Ge scholar, reached the highest rank possible amongst ministers. To embrace the prestige and fame, Chen family added another circle of tall walls in front of the Ming Dynasty castle. The additional new walls included two of the fame Arch and the official courtyards, thus formed a magnificent castle of the Twin Cities. The part of the Ming Dynasty architecture named "Douzhuju", the other part built in the Qing Dynasty called "Zhongdaozhuang," Huangcheng Xiangfu is the modern name used in the recent development of tourism.

Whether taking account of the heritage values of recorded historical and cultural information, or considering the quality of construction, Huangcheng Xiangfu should be the leader in the existing ancient castles.

First, it mostly reflects the classic characteristics of ancient castle buildings.

Huangcheng Xiangfu was built in the mountains, various inner buildings located in the layers of artificial trimmed terraces, and walls built along the terrains of the mountain. In overview, layers of horizontal structure integrated by the slash lines of the walls, also interwoven with the vertical lines of the Heshan Tower, make wonderful outlining contours as well as showing the endless details; grand and magnificent intertwined with fine picture of the pastoral contented life seamlessly. Heshan Tower and many turrets in walls concisely expressed the characteristics of the Wubi construction. Especially Heshan Tower reaches 28 meters high with squared external shape, the wall actually has thick lower part and thin upper part. Therefore, stronger strength in the lower part coupled with greater interior space in the upper part, both beautiful and practical in deed. Within the building there are also wells and underground passages. Rear walls combined with the platforms to form 3-layer cave-shaped hiding place for soldier. The contrast between the arc and heavy mass of the cave, and airy and light weight wooden structure building, greatly enhance the artistic characters of the form/shape. Such not only dignified and flexible but also solemn and elegant design, was based from a strong farming culture in Shanxi, meanwhile, from the contribution of nomadic culture and mercantilist culture in the region as well.

Second, the condensed record of the status of the ancient pastoral society.

The ancient Chinese society has been pursuing agriculture-oriented culture, but the real wealthy pastoral were fundamentally due to the setting up of industry and commerce, traditional ethics and the contradictions of reality laid the bane of social conflict. In the late Ming Dynasty, corruptions spread from top to bottom of the society, industry and commerce has also ushered in a period of prosperity, which may be linked to the government officials

的艺术性。如此既端庄又灵活、既厚重又飞扬的设计,一方面来源于山西深厚的农耕文化底蕴,另一方面也有此地的游牧文化和重商主义文化的贡献。

第二,集中记录下古代田园社会的状况。

中国古代社会一直奉行以农为本,但真正富裕的田园都是因兴办了工商业,伦理与现实的矛盾从根本上埋下了社会冲突的祸根。明代中后期,社会自上而下腐败的同时,工商业也迎来了一次繁荣期,这可能与政府官员忙于党争而放松了对社会的压制有关。但因为没有制度配合,社会上贫富悬殊现象、浮躁现象、全民经商现象越来越严重,光靠田园社会自身是解决不了这些问题的。当时的沁河一带肯定也有贫富之分,但由于总体上比较富裕,区域也不大,矛盾就可以化解,但区域一大就不行了。当陕西出现大规模饥荒时,社会无粮赈济,必激起民变,从此恶性循环,民变规模越来越大。民间的自保团体这时必须有相当的人力物力才能有效自保,简易的寨堡很容易被摧毁。沁河的富商有经济实力,可以修坚固堡垒,同时,这里又必须有平日在农耕文化的气氛中形成的相对融洽的乡里关系,有商业文化中形成的合作关系,紧急时刻大家才能联合起来。所以,一个社会必须农工商并重,有自主的民间关系的发育才能稳定、才有力量,这是如皇城相府般的沁河古堡相比其他地方更能清晰地提示给人们的东西。

从河山楼方向望郭峪的豫楼　The outlook view of Yu Tower from Heshan Tower in Guoyu castle.

皇城相府平面图
The plan of Huangcheng Xiangfu.

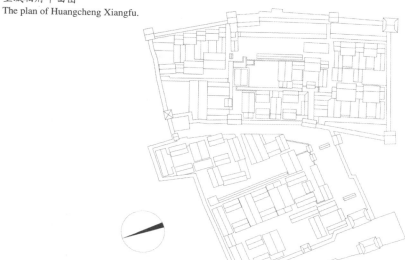

河山楼正面影像和立面图（意大利马方济测绘，2012.3.）
The front image and elevation drawings of Heshan Tower (Survey by Italian Prof. Francesco Maglioccola, March, 2012).

在"中道庄"的城墙上望"斗筑居"
View of Douzhuju gate from the city wall of castle.

"斗筑居"后部，体现出山地古堡建筑的特征
The rear section of Douzhuju, demonstrates the characteristics of mountain castles.

第三，展现田园文化对中国文化的贡献。

首先在建筑方面，以民间微薄之力，山西的民间建筑常令呆板的官方建筑相形见绌，至于古堡，情况基本上是民间古堡填补了中国古建筑的这一空白。

更重要的是文化建设，尽管古堡在欧洲的文化中占有很重要的位置，但总体上，欧洲的古代文化主要是由城市孕育，中国古代文化则主要由田园孕育，中国古代的读书人几乎都在田园中受教育，即使后来进城为官、为商，最终的归宿还是在田园，田园是中国文化的根，古堡常常在危机时能保护住这个根。

were busy with the party struggles and relaxed on social repression. However, because no system to regulate the society, the phenomena of disparity between the rich and the poor, impetuous within the community, and nationwide focus on trading/business became more and more serious. Rely on pastoral community itself can not solve these problems. Then along the Qinhe River region certainly had rich and poor separation, because the area was not large and generally more affluent, the contradiction could be resolved, however, it was definitely not applicable for a large region. There was no food relief for the society during Shaanxi large-scale famine, civil commotions were stirred up and formed a vicious cycle of increasing scale of civil disorder. Non-governmental self-protection groups then must have considerable manpower and material resources in order to effectively protect themselves, because simply built stockade village was very easy being destroyed. Qinhe's wealthy rich had economic strength and was able to build sturdy castle. At the same time, there must have a relatively harmonious Township inherited from the farming culture, plus cooperative relations formed in the business culture, so that in times of emergency community efforts can be joint up together. Therefore, a society must have harmonic development of agriculture, industry and commerce, the independent non-governmental relations can be stable and powerful. Huangchen Xiangfu representing Qinhe castles clearly showcased the important lessons for a sustainable community in ancient era.

Third, showcase the contribution of the pastoral culture to Chinese culture.

First of all, Shanxi's civil modest vernacular architecture often distinguished and outperformed the dull official buildings. In terms of castles, private castles basically fill in the gap amongst the ancient Chinese architecture. More important is its contribution to Chinese culture. Construction of castle occupies a very important position in European culture, in general, Europe's ancient culture mainly incubated in the city. Ancient Chinese culture was nurtured by the agricultural pastoral. Almost all ancient Chinese scholars were educated in pastorals, even though they advanced into cities as governmental officials, or commence providers, their final destinations are still back to the countryside. Rural pastorals are Chinese cultural roots, castles often were able to protect these roots in times of crisis.

从西侧俯瞰皇城相府
The bird's eye view of Huangcheng Xiangfu from the west.

6. 沁水县湘峪堡

沁水古堡群涉及的明代名人非常多，陈廷敬的夫人是明代著名财经专家王国光的后人，王国光著有财经名著《万历会计录》，协助张居正进行了中国历史上一次著名的改革。王国光是与皇城相府一山之隔的上庄村人，那里虽然没有古堡，但那里的堡院特别突出看家楼。

湘峪堡与皇城相府也是一山之隔，那里是明朝另两位知名大臣孙居相、孙鼎相兄弟的故里，此兄弟二人均入选魏忠贤阉党给政敌编的《东林党点将录》，分别为其中的"天暴星两头蛇兵部右侍郎孙居相，天哭星双尾蝎左副都御史孙鼎相"，说明二人不肯与阉党同流合污，固守田园文化的核心价值。

湘峪堡同样位于沁河的一条支流边，虽然城墙是临河而立，不过现在古堡前面的大水面是近年筑堤蓄出来的，这样为古堡增添了戏剧性，但也降低了城墙的高峻感。与皇城相府一样，当年这些古堡前面都是漫滩，没有现在被填高的公路和广场。

临河的城墙宽约 300 米，古堡中多为两三层楼房的民居鳞次栉比，更有塔楼林立，这些高塔有堡门、角楼、碉楼、插花楼等，还有西侧紧邻的东岳庙的钟鼓楼，共有十四五座，如此景象，实可谓壮观。进堡后又可以看到，紧挨城

俯瞰湘峪堡　　The bird's eye view of Xiangyubu Castle.

6. Xiangyubu Castle in Qinshui County

The Qinshui Castle group involved many celebrities in the Ming Dynasty. Chen Tingjing's wife is the descendants of the famous Financial experts Wang Guoguang in Ming Dynasty. Wang Guoguang wrote a financial masterpiece the *"Wanli accounting records"* to help Zhang Juzheng to conduct a famous reform in the Chinese history. Wang Guoguang's village is a mountain away from the Huangcheng Xiangfu, where there is no castle, but particularly prominent of the Fort courtyard and the watching tower.

Xiangyubu Castle is also a mountain away from Huangcheng Xiangfu, it is the hometown of two well-known ministers of the Ming Dynasty, the Sun brothers – Sun Juxiang and Sun dingxiang.

Xiangyubu Castle is also located near a tributary of the Qinhe River, the original walls were still standing on the riverbank, but the large water body in front of the castle only formed by building embankment in recent years. This water adds a dramatic effect for the castle, but reduces the steep and stiff senses of the walls. Similar to Huangcheng Xiangfu, there was only floodplain in front of these castles in ancient time, not filled by the elevated roads and squares in present time.

The walls on riverbank has a width of 300 meters, inside the castle many two- and three-storey houses stand row upon row, and more towers intersect in between. These towers

俯瞰湘峪堡　　The perspective view of Xiangyubu Castle.

湘峪堡右侧城墙，城墙后的楼房相当于二层城墙
The right-hand-side of the city wall of Xiangyubu Castle, the houses behind the wall are equivalent to the second layer of the city wall.

法国卡尔卡松城堡的双层城墙
The double-layer city wall of Carcassonne Castle, France.

include gate, turrets, towers, watching towers, plus the Dongyue Temple's Bell and Drum Tower adjacent to west side, a total of fourteen to fifteen buildings, such a scene can definitely be described as spectacular. Stepping into the castle, you can see that the lower part of many buildings next to the walls have hiding place for soldiers, the upper windows can be used to discharge arrows and fire shots, it in fact forms the double-layer walls. Counting the facing-out windows from the hiding place for soldiers in the outer-layer of walls, plus the bridleways on the top of the wall, form a total of 3 layers outwards attacking platforms. This structure is reminiscent of the famous Carcassonne Castle in France. The excellent defensive settings enabled Xiangyubu Castle to withstand the next two attacks by the Shaanxi hungry.

3 图：湘峪堡正面左侧城墙，在堡内地平面下，城墙中还有一层防御空间
Three above: the left-hand-side of the city wall of Xiangyubu Castle, there is hidden layer of defence space under the castle's ground level.

墙的许多楼房下层都是藏兵洞,上层的窗户可以对外放箭、射击,事实上形成了内外两重城墙。算上外重城墙中藏兵洞的向外窗口、墙顶上的马道,共有3层向外攻击的作业面,这种构造让人想起了法国著名的卡尔卡松城堡,出色的防御性使湘峪堡至少抵挡住了陕西饥民随后的两次攻击。

在古堡内的主台地上,除了南门楼前有一小块宽敞处外,堡内全是窄巷,窄巷上空有许多过街楼,看来湘峪人把巷战的准备也做好了。

当年在孙氏兄弟的倡议下,湘峪堡在崇祯七年建成。现在堡内最显眼的5层大碉楼就在孙鼎相的故宅内,这座楼与河山楼、豫楼不同,它是宅院的正房,沁河的宅院中常见3层高的实体型正房,而这座正房因为太高又貌似碉楼,反映出设计上的灵活性。孙居相的宅院损毁严重,万历赐建的石牌坊也只剩残柱。

3图:右侧双层城墙间的巷道
Three above: Tunnels between the double-layer city wall on the right side of Xiangyubu Castle.

湘峪堡内插花楼院落的正面
The front view of the courtyard with watchtower in Xiangyubu Castle.

插花楼院落的背面
The rear view of the courtyard with watchtower.

下2图：孙鼎相故宅内的碉楼式正房
Two below: The main Diaolou house in the previous residence of Sun Dingxiang.

孙鼎相故宅边的污水池和水牢
Waste water pond and water prison next to the previous residence of Sun Dingxiang.

孙氏兄弟古宅中的石雕
Stone carvings in the Sun brother's residence.

湘峪堡内街巷中的过街楼
Over street bridge-house in Xiangyubu Castle.

皇城相府和湘峪堡之间的设防村落
The walled village located between Huangcheng Xiangfu and Xiangyubu Castle.

Within the castle, except a small open area on the ground of the main platform in front of the south gate, narrow alleys are everywhere, many bridge galleries are placed over the alleys. It seems that Xiangyubu people have done the preparation for street fighting as well. Under the initiative of the Sun brothers, the building of Xiangyubu Castle was completed in the 7th year in Chongzhen era. Now the most prominent five-storey tower is located inside the Sun Dingxiang's residence. This tower is different from the Heshan Tower or Yu Tower, it is the main house in the courtyard. Normally a three-storey main house is more common in Qinhe residences. This main house is extremely tall and looks like the Diaolou, that reflecting the flexibility in design. Sun Juxiang's house was severely damaged, the stone archway honourably given by the Wanli Emperor only has residual pillars.

和山西所有古堡一样，湘峪堡有复杂的地道系统，村中老人在讲述抗日战争时期她们躲在地道中的往事，地面上的大块石板下就是地道的出入口或通气孔
Similar to all castles in Shanxi, Xiangyubu Castle has complicated underground passage ways. The elders in the village are talking about the past when they were hiding inside the underground tunnels during the anti-Japanese war time. The entrances/exits or air vent of the tunnels are beneath the giant stone slates.

7. 沁水县柳家堡

唐代大文豪柳宗元是山西永济人，据说在他因搞改革而遭到人身攻击后，其族人至少有一支迁入中条山，落脚在现在历山脚下的西文兴村，繁衍至明代，成为当地巨族，在陆陆续续的建设下，在沁河主流西面的一条支流边，形成了一座古堡式大宅。

古堡位于一座高台上，周边垒起陡峭的护坡，加上高大外墙，便形成了极好的防御性，高台前后只有两条坡道供上下，坡道尽端都有高大门楼，内部也有地道、暗墙。柳家堡内外府的格局颇有中国古代大型官式建筑前庭后苑、前署后宅的气势，外府的空间大多是献给各路神灵的，有宗祠、虞帝庙、文庙、关帝庙等。内府原有13座宅院，现存7座。

作为文豪之后，柳家堡内藏有朱熹、王阳明、文征明、王国光等名人的几十通书法碑，俨然有座私家碑林，还有文昌阁、魁星楼，两座高楼都与古堡的防御功能结合起来，魁星楼实际上充当了堡门，文昌阁守卫着内府。

柳家堡入口
The entrance of Liujiabu Castle.

堡内巷道
The roadway inside the castle.

兼作一道堡门的文昌阁
Wenchang Temple is also used as a gate.

7. Liujiabu Castle in Qinshui County

Tang Dynasty great writer Liu Zongyuan is from Yongji in Shanxi Province. It was said that he was attacked after engaging in the reform of governing system, at least one of his family groups moved to Mountain Zhongtiao. This group settled in Western Wenxing Village at the foot of Moutain Lishan, grew and expanded in the Ming Dynasty and has become a giant rich family. After continuous construction, a castle-style mansion was built beside a tributary in the west side of Qinhe mainstream.

The castle was located on a high platform, with built-up steep slopes in the surroundings, plus tall external wall, it formed an excellent defensive castle. In front of and behind the platform, there were only two high-ramp paths for up and down connections. The ramp reaching a tall gatehouse at the end, inside the gate there were underground passages and hidden walls. The inner and outer houses of Liujiabu Castle mimicked the governmental pattern of large Bureaucracy building of ancient Chinese, with vestibule for official duty and back courtyard for domestic residence. The outer house space is mostly dedicated to ancient legends and spirits, including ancestral hall, Yudi Temple, Confucian Temple, Guandi Temple, etc. The original 13 houses within the inner housing compound, only seven exist at present.

As the residence of a great writer's later generations, Liujiabu Castle kept dozens of Zhu Xi, Wang Yangming, Wen Zhengming, Wang Guoguang and other celebrities' calligraphy monuments, apparent private collection of steles. The two buildings, Wenchang Pavilion and Kuixing Tower, together with the castle combined the defense functions as a whole, the Kuixing Tower actually acted as a gatehouse, Wenchang Pavilion, guarded the inner house.

右2图：堡内堡院的两座门楼
Two right: two gates of the courtyards inside the castle.

8. 阳城县寨山古堡

阳城县北留镇南岭村西坡庄的正东面，一片巨大的、颇有世外桃源境界的河谷中间耸立着一座小山，山的东南面壁立千仞，沁河的支流长河从山壁下流过，山的西北面也是陡坡，一圈圈的岩石沿小山的等高线裸露出来，使山顶上的一座石头的小古堡更加像是从石头山上长出来的，而等高线状的裸露石头像是古堡的外层城墙。肯定是因为有小古堡的存在，那座小山被称为"寨山"。

关于寨山古堡的历史已无从考证，民间传说春秋时期在鲁国爆发的柳下跖奴隶起义军曾经来过这里，山下河边处的一座小山村名为"石堂"，相传是柳下跖当年埋锅造饭的地方。古堡的历史估计不会这么早，其内部的关帝庙中有明代的字迹，古堡的建筑历史应该更早。

古堡呈不规则圆形，仅在南面设一门。城墙高十余米，厚1米多，全部用片石垒砌，上部有雉堞遗迹。堡内除了有一座关帝庙，还有据说是以八卦图布局的、片石垒的房舍，当地幼童在残垣断壁间玩耍，常常会被困在里面，找不到出口。

与前述的沁河古堡不同，寨山古堡应该是早于明末大商户所建古堡的一种更原生性的古堡，它可能是周边村民常设的避难所，也可能是曾经的上山为王、落草为寇者留下的。（王文彬撰文）

远望古堡　Outlook of the castle on the hill.

8. Zhaishan Castle in Yangcheng County

In the east of Xipo village in Nanling village of Beiliu town in Yangcheng County, a mountain stands in a vast valley of paradise realm. A steep cliff formed in the southeast side of the mountain, Changhe River (one of Qinhe's tributaries) from streams passing the cliff, the northwest side of the mountain is also a steep hill. Rings of rocks popped out along the contour lines of the hill, a stone castle on the hilltop is more like growing out of the Stone Mountain, while the contour-lined bare stones are just like the outer wall of the castle. The hill known as the "Castle Hill", must be because of the existence of the small castle.

It is hardly to trace back the history of the Castle Hill (Zhaishan Castle), according to the folk legend, Liu Xiazhi's slaves army had come here during the outbreak of rebellion in the Chunqiu era in Lu Kingdom. A small mountain village beside the river at the foot of the mountain is known as the "Shitang", where was the cooking place for Liu Xiazhi's army. The history of the castle is not expected to date back in Chunqiu era, but the writing inside the Guandi Temple proved that the architectural history of the castle should be dated earlier than the Ming Dynasty.

The castle has irregular circular shape, there is only one gate located in the south. External walls are more than ten meters in height and 1 meter thick, all constructed with slat stones, and the remains of battlements are in the upper part. Inside the castle, in addition to a Guandi temple, it was said to have slat stone built houses in the layout of the Eight Diagrams pattern, the local children often trapped in the ruins while playing, couldn't find the exit.

Distinct from aforementioned Qinhe castle, the Zhaishan castle should be a more native castle constructed earlier than the castles built by rich merchants in the late Ming Dynasty. It might be permanent refuge for the surrounding villagers, but also might be the fortification of former rebellion kings and local rulers. (written by Wang Wenbin)

古堡内部
Inside the castle.

古堡后部直对河谷　The rear of the castle facing the river valley straight on.

9. 泽州县"玫瑰堂"古堡

现泽州县的大箕镇位于古代的晋豫古道边,为当时古道上繁华的商镇,镇边一片谷地上,一座小山兀然孤立,在山顶可俯瞰古道。不知何时,山顶上长出了一座古堡,这种姿态颇似欧洲中世纪那种以收"过路费"为目的兴建的古堡。也许正是因此,后来古堡里建起一座教堂,这种古堡与教堂的组合在欧洲比较多见,在中国恐怕绝无仅有。

名为圣母玫瑰堂的教堂是 1914 年由荷兰传教士建立的,而从古堡墙体的质感看,它所经的岁月远不止一百年。古堡呈不规则椭圆形,堡门右侧的碉楼与古堡主体呈 45°角布局,这与许多汉代明器的形式相仿。古堡更多的历史已无据可查,只知此地有个地名叫小寨,或许是因这小古堡而得名,因为古代将堡常称为寨。

"玫瑰堂"古堡正面　The front view of Rose Hall Castle.

9. "Rose Hall" Castle in Zezhou County

Present Daji town in Zezhou county is located along the ancient trail between Shanxi and Henan, it was a bustling commercial town on the trail in ancient time. Next to the township above a field, a solitary hill stood alone overlooking the trail from the hilltop. Unwittingly a castle appeared on the hill top, it very much resembled the purpose-built "toll" castle in the medieval Europe. Later on, a church was built within the castle, this combination of castle and church was more common in Europe, probably very unique in China.

The church called Our Lady of the Rosary Church, established in 1914 by the Dutch missionaries. According to the castle wall texture, it existed far more than a hundred years. The castle shaped as an irregular oval, the tower on the right side of the gate combined with the castle main body to form a 45-degree layout, which is similar to the forms of many funerary objects in the Han Dynasty. More historic details of the castle are no longer documented. The only clue was linked to a nearby place being named as Xiaozhai, most likely it was named after this small castle, because ancient Castle is often referred to as Zhai.

古堡的雄堞
Castle battlements.

古堡侧面 Side view of the castle.

10. 河南省博爱县寨卜昌村

沁河穿过太行山后进入河南的平原地区，小型城堡不再有很好的防御效果，但在一些特殊时期还是有设防村落形成，其中有个别的保留了下来。

在沁河北岸的博爱县一座名为寨卜昌的村落中，村民以王姓为主，其先祖在明代初年的大移民时从山西洪洞县迁到此地，于清代因销售山西出产的铁锅而致富，估计当年他们的客户中就有樊溪陈家、王家等。与明代后期的情况一样，由于社会矛盾激化，清代中后期分别爆发了太平天国和捻军起义，其中捻军的主要活动区域包括河南，王氏家族未雨绸缪，发动乡民筑起了一道城墙将村落保护起来。1949年后，因为城墙是封建的象征，有人曾经组织了几千人要把城墙铲平，但由于城墙的夯土过于坚硬，铲了一部分后，人困马乏，只能作罢。这也是因为那时候缺少机械，城墙才能幸存下来一部分。作为一座村落，这里的城墙尺度与一座县城的差不多，非常少见。

与欧洲古堡古城的城墙多用石头不同，中国多用夯土，夯土虽然不如石头永恒，但有些夯土由于与石灰等材料配比恰当，夯实过程严谨，也能经受千百年的风雨。由于曾经作为"四旧"，古堡特别是它们的城墙都属于必须铲平的东西，古堡的建筑能够保留下来，主要是因为当时都有人住，城墙能保留下来，主要是因为它们太坚硬。

夯土城墙　City wall made of rammed earth.

10. Zhaibuchang Village in Bo'ai County, Henan Province

Qinhe River runs through the Taihang Mountains entering into the plains of Henan, a small castle no longer has a good defense function. However, in some special periods there were fortified villages formed, just few of them retained.

In a Zhaibuchang village in Bo'ai county at the north bank of Qinhe River, most of the villagers surnamed Wang. During a large-scale migration in early Ming Dynasty its ancestors moved here from Hongdong county in Shanxi. In the Qing Dynasty, the villagers got rich by sales of Shanxi iron woks, probably their clients included Fanxi's Chen family and Wang family, etc. Same as in the late Ming Dynasty, social conflicts approached peak time in the late Qing Dynasty period, there were outbreaks of the Taiping and the Nian Army Uprising. Due to the Nian army's main activity area including Henan province, Wang family planned ahead and mobilized villagers to build a wall to protect the village. After 1949, because the walls are a symbol of feudal, it was attempted to organise thousands of people to put the walls down. But the rammed earth walls were too hard to destroy, after shoveling a small part, exhausted crowd could only give up. This is also because the lack of machinery at that time, part of the walls has luckily being preserved. As a village, the walls' scale almost matched with the scale of a county that was very rare.

Walls of European castle were mostly built by stone, very differently, many walls of Chinese ancient cities were built by rammed earth. Although rammed earth is not perpetual as stone, but some of rammed earth mixing with lime and other materials in appropriate ratio, plus rigorous compaction process, then can withstand thousands of years of wind and rain. Had regarded as the "Four olds", the castle, in particular, their walls are necessary to be levelled out. Castle building can be preserved, mainly because there were people living in; walls can be preserved, mainly because they are too tough to remove.

村内的民居
Houses in the village.

村中住宅建筑的装饰细部　The architectural details in the houses in the village.

四、汾河流域民间古堡

汾河是黄河第二大支流，也是山西最重要的河流，它发源于山西北部，一路南流，经太原、平遥、介休、灵石，到侯马后调头向西，在河津注入黄河。汾河沿线集中了山西最多的古城，从前这些古城都有城墙，现在只有平遥完整地保留下来。除了古城，汾河流域也有大量古堡，位于深山中的古堡可能是临时避难所，现在只剩残墙，而平原、丘陵地区的正常村落很多也是设防的堡村，或是里坊制的堡村。现存的汾河古堡多为清代建筑，比沁河古堡历史要晚。与沁河古堡一样，汾河古堡也多与晋商有关，清代的晋商以晋中人为主，虽然晋中的煤炭业和冶铁业并不发达，但贸易和金融业可以创造更多的财富，使这里的古堡更为华丽。

Section 4 Vernacular Castle in the Fenhe River's Basin

Fenhe River is the second largest tributary of the Yellow River, also Shanxi's most important river, which originates in the northern Shanxi Province, and flows southwards all the way. Passing Taiyuan, Pingyao, Jiexiu to reach Houma, then after a U-turn at Houma it began to flow to the west into the Yellow River at Hejin town. Most of ancient cities in Shanxi are gathered along the Fenhe River. In the past, all of these ancient cities have walls. At present, only Pingyao has completely preserved walls. In addition to the ancient cities, there are also a large number of castles in the Fenhe River Basin. Castles in the mountains were temporary shelters, only broken walls left. Normal villages in the plains or hilly areas were also walled village or walled Li-fang system village. Existing Fenhe River castles were mainly built during the Qing Dynasty architecture, later than Qinhe River castles chronologically. Similar to the Qinhe castle, the Fenhe Castle was also linked to merchants mostly from Jinzhong in the Qing Dynasty. Coal industry and the metal fabrication industry were not well developed in Jinzhong, but trade and financial sector could create more wealth, so that more magnificent castles were built here.

1. 平遥县的堡村

对于许多平遥古城周边的乡民来讲，平遥这个大堡垒只是将他们一直居住的小堡垒放大了十几倍而已，如果人们想在了解古代城市后再了解一下这里的古代乡村，可去平遥古城东南面一些名为××壁的古堡形村落。

走入乡野后，可见村落都分布在起伏的黄土地的高台地上，被树林簇拥着，看得出所有的老村从前都是堡村，有防御性的城墙。岳壁的堡墙轮廓完整，局部还残留雉堞，内部格局也很完整，丁字形主街上分布着几座小神庙，横竖街交叉处是村内李氏家族的宗祠，这是堡村一种典型格局。

岳壁东南2公里的梁村是由6个独立的堡挤在一起形成的大村，每个堡都有高大的堡门，门前的巨树标志着古堡的久远历史。堡门为2层，上层供保护神像，门内主巷道两侧也多为2层高的院墙，一般在离堡门很远的地方才开院门，且院门很少，这有增强防御性的考虑，也有古代里坊制度的遗痕，恐怕还有风水上的讲究。

虽然没有乔家大院等那些著名的晋商大院豪华，但堡村内也不乏精美的院落，只是每个院落分属于各个中产阶级家庭，互不贯通。堡村有自己独特的美感，除了自然古朴以外，它们有古堡特有的一种静穆，厚重单纯的砖墙因风雨侵蚀而表情丰富，封闭压抑的空间因战火消融而珍存住了灵气。

继续向东南走还有赵壁、段村等堡村，赵壁古堡位于一座土丘上，入口处的神庙一直盖至土丘顶，为古堡构筑出抑扬顿挫的造型。段村的堡很多，其中几座内有比皇城相府河山楼细许多的碉楼，说明其所属家族的人口比较少。

堡村一般建在高地上，其中一些有高起的碉楼
Walled villages were normally built on high ground, some have Diaolou.

梁村南乾堡的堡门
The gate of Nanqian castle in Liangcun village.

1. Castle villages in Pingyao County

In the view point of surrounding villagers, the ancient city of Pingyao being a large fort, was equivalent to the small fortress they have been living magnified more than ten times. If people want to understand the ancient city and then look at the ancient villages, they could go to the southeast of Pingyao and visit the castle-styled villages called certain Bi.

Into the countryside, showing that the villages are located in the loess plateau of undulating high platforms, surrounded by woods. It can be seen that all the old villages were formerly castle villages with defensive walls. The castle wall outline of Yuebi is complete, local residues battlements, the interior layout is also very neat. Several small temples distributed on the T-shaped main street, Li family's ancestral hall located at the intersection of the streets, which is a typical pattern of a castle village.

Liangcun village, about 2 km southeast of the Yuebi, has 6 castles jointed together to form a large village. Each fort has a tall gate, the giant trees in front of the gates mark the castle's long history. The gate has two layers, the upper layer for worshiping protection statues. Inside the door on both sides of the main roadway are mostly two-storey walls, normally very far from the gate there is a door open to the courtyard. Very few doors, in addition to enhanced defensive considerations, plus the traces of the ancient neighborhood system Li-fang, there might have particular interpretation in *feng shui* as well.

Although there is no famous Shanxi merchants' grand compounds, e.g. Qiao Family's luxury courtyard, but there is no lack of beautiful and elegant courtyards. Each courtyard belonging to various middle-class families, and does not link up. Castle village has its own

梁村天顺堡堡门内部
The inside view of the gate of
Tianshun castle in Liangcun village.

堡村内的晋商住宅
Shanxi merchant's house in walled village.

2. 介休市张壁

张壁位于介休市东面的黄土丘壑之间，据记载，它起于隋末，当时群雄并起，后来的唐高祖李渊从山西打入关中，刘武周则从朔方南下掏了李渊老巢。后来的唐太宗李世民急忙回救，先在晋南新绛一带的柏壁击退刘军，又在介休歼灭刘军主力。

据说当年刘军还有一支部队藏在张壁，战争过后，这批人就在张壁住了下来。

我们现在看到的张壁是否就是隋末的张壁已不得而知，但这段故事估计是真的，因为刘武周不仅是个失败者，他还投奔突厥，在很多人的固有观念中这属于叛国投敌，张壁人要不是真的与他有特殊关系，谁愿跟他有瓜葛，还建了座华丽的可汗祠祭奠他，称他"生为夷狄君，殁为夷狄神"。

张壁古堡远看很低平，近看才知它三面都有深沟，只有南门前地势较平，也有小河流过，河水向西汇入汾河。南门右侧一座高台上布满建筑，其一半突出堡外，像是南门的瓮城，一半嵌在堡内，台上的建筑是关帝庙、魁星阁、可汗祠。南门内是堡内主街，向北对着另一座高台，上面的建筑有真武庙、吕祖堂、二郎庙等，两座高台都犹如堡中之堡。主街上还有祠堂和其他寺庙，中国古代城镇的构造中心往往由神灵空间统筹，但像张壁这样强烈、隆重的神灵轴线还不多见。

一般来讲，坐北朝南的中国北方古建筑群体都是南低北高，但张壁是北低南高，人从古堡正面进入堡内后越走越低的感觉很强烈，不知这与张壁地下乾

张壁正门
The main gate of Zhangbi Castle.

unique beauty, in addition to natural simplicity, a unique serenity of ancient castle, wind and rain erosion added rich texture on the simple thick wall. The closed and oppressive space faded away with the war, then replaced by cherish living aura.

Continue to the southeast, there are castle villages, e.g. Zhaobi and Duancun. Zhaobi castle is located on a hillock. The temple at the entrance has reached the hilltop, forming the stereo dynamic shape of the castle. There are many castles in Duancun village, including several towers much slender than the one in Huangcheng Xiangfu-Heshan Tower, which reflect smaller population in the family group.

2. Zhangbi Castle in Jiexiu City

Zhangbi Castle located in the loess hills and gullies in the east of Jiexiu City, according to historical records, it started from the late period of Sui Dynasty, while all states contending for hegemony. Later, the first Tang Emperor Li Yuan invaded Guanzhong from Shanxi. In the meanwhile, Liu Wuzhou advanced south from Shuofang and destroyed Li Yuan's lair. The second Tang Emperor Li Shimin hurried back to the rescue, winning the first battle at Bobi in vicinity of Xinjiang region in the south of Shanxi and, drove Liu's army back, then annihilated the main army in Jiexiu.

It was said that there was a branch of Liu's army hidden in Zhangbi, after the war, these people settled down in Zhangbi.

It is not known whether the Zhangbi we see now is actually the Zhangbi since Sui Dynasty, but the story is probably true, because not only Liu Wuzhou was a loser, he defected to the Turks. According to the concepts inherent in a lot of people, this was treason. Zhangbi people either really had a special relationship with him, no one would be willing to keep the connections. They also built a splendid Khan Temple to pay homage to him, saying he was live as a barbarian monarch, die as a barbarians' god.

Viewing from a distance, Zhangbi castle appeared as very low-lying structure. Taking a closer look, it actually has deep grooves on three sides. Only the the terrain front of the south gate is relatively flat, there are small rivers flowing west joins Fenhe River. A high platform on the right side of south gate is covered with buildings, half of them exposed outside the Fort, such as the south gate barbican, half embedded in the castle. The buildings on the platform are Guandi Temple, Kuixing Pavilion, and Khan Temple. Within the South gate is inner main street, north facing another set of buildings on high-level platform, i.e. Zhenwu Temple, Luzu Temple, Erlang Temple. The two high-level platforms (building sets) appear like castle within castle. Temples are also placed on the main street, the structure of the ancient town center is often guarded by the gods/spirits space, but it is still unusual to

坤的吸力有无关系。张壁虽有高墙深壑，但其更重要的防御系统是地道，地道有3层，是个庞大的立体网络，地面上用各种隐蔽方式连通每座庙宇民居，也有出口通向堡外，地道总长近万米，最深处20余米。

有建筑学家认为，游牧人建城或建堡爱采圆形，是源于他们围羊圈的习惯；农耕人建城或建堡爱采方形，是源于阡陌的形态。我们看到张壁是圆形的，我们也知道刘武周的军队中有不少游牧的突厥人。不知这两个现象之间有无逻辑关系。

历次中原战乱时，中原汉人并非只向南跑，不少人是向北跑的。"隋末乱离，中国人归之者无数，遂大强盛，势陵中夏。"(《隋书·列传第四十九·北狄、突厥》)就是说隋末突厥之所以强盛是因大量中原人逃入其中，山西张壁古堡那里就表现了这一点。

上高台的坡道
The ramp to reach the high platform.

堡内主街后段上的古树，其旁有水池
An ancient tree at the rear section of the main street in the castle. There is a pond next to it.

emphasize the grand axis of the gods so strongly in Zhangbi.

Generally speaking, the ancient building groups facing south in northern China are low in the southern part and high in northern part, but Zhangbi is higher in the south than the north portion. Entering the castle from the Zhangbi's main gate, the sense of gradually stepping down is very strong, it might be linked to the suction of Zhangbi's underground world. Although Zhangbi has tall walls and deep valleys, the more important defense system is the underground passages. The passage system has three layers, that is a huge three-dimensional network. On the ground with all kinds of hidden connectivity of each block of the temple or house, this system also has exits leading to the outside of the castle. The total length of the passage way is nearly 10,000 meters, reaching down to 20 meters at the deepest section.

The architects believe that the nomadic people love to adopt a circular shape to build a city or a castle, which is derived from the habit of sheep penning; farming people love to adopt a square shape for city or castle building that is derived from the form of terraced rice paddies. Zhangbi is a circular castle, we also know that there are nomadic Turkic people in the army of Liu Wuzhou. Not sure whether there is a logical relationship between these two phenomena.

古堡后部的城墙
The rear section of the castle wall.

古堡内空中的连接系统
The inter-connecting system of the castle.

古堡内丰富的空间
Versatile spatial design within the castle.

张壁古堡主体构造示意图　Illustration of the main architectural layout of Zhangbi Castle.

堡内后部高台上的真武庙
The Zhenwu Temple on the terrace at the rear of the castle.

高台侧面下部
Details of the lower part beside the high platform.

堡内巷道
Roadways in the castle.

3. 灵石市恒贞堡

在所谓的晋商大院中，位于乡间的几乎都是古堡式院落，而如灵石县静升镇王家大院，则完全是古堡。实际上，王家大院是由几座名称即为××堡的古堡群落组成，现在保存完好的是红门堡，或称恒贞堡。

王家的发迹也是靠商官结合，在清代乾隆至嘉庆年间，家族势力达到顶峰。红门堡于乾隆年间兴建，那时候据说是盛世，虽说盗匪不会一点儿都没有，但王家兴建如此的大堡，主要目的不应是为了安全，而是为了体现家族势力。因此采用城堡形态，一方面是因为山西民间本身有建堡的传统，城堡的形象高大威武；另一方面是中国的门阀士族自古有建坞壁的传统。

为了同样的目的，王家的城堡在布局上追求的首先是礼制，城堡整体呈规整的长方形，外形完全对称，内部布局也大体对称，强调等级、秩序和里坊格局，相比沁河的设防村镇，红门堡更有城市意象，看来王家向往家宅城市化。

从另一座古堡"高家崖"连通恒贞堡的桥。高家崖没有高大城墙，其防御性主要靠它地处高台，近似宁夏的台式建筑
The bridge connecting Gaojiaya to Hengzhenbu. Gaojiaya has no tall castle walls, it's located on the high-level terrace for defence function, which is very similar to the terraced buildings in Ningxia.

3. Hengzhenbu Castle in Lingshi City

Almost all the so-called merchants compound in the countryside are castle style courtyards. For example, the Wang family's house in Jingsheng town, Lingshi County, is definitely a castle. In fact, Wang's house is a castle group composed of several named certain castles. To date, the well preserved one is Hongmen Castle or Hengzhenbu Castle.

Wang family's fortune relies on the close collaboration of merchants and officials, the family influence peaked during Emperor Qianlong to Jiaqing years in the Qing Dynasty. Hongmen Castle built in the reign of Emperor Qianlong, that time is said to be peace and prosperity. Although the bandits will not vanish completely, the main purpose of Wang family built this grand scale fort should not be for security, but to reflect and display the family's power and influence. Therefore, the castle form was chosen because of the majestic image of the castle, in line with Shanxi folk tradition of castle construction; on the other hand, the aristocratic gentry and clan of China, have the tradition of building Wubi since ancient times

For the same purpose, the Wang's castle firstly was on the pursuit of ritual system in the layout. The castle as a whole was neat rectangular shape perfectly symmetrical, the internal layout was also largely symmetrical pattern to emphasis on grades, orders and Li-fang system. Compared to the fortified village of Qinhe River, Hongmen Castle appears a more urban image, which indicates Wang family was longing for the urbanization of residence.

远望恒贞堡　　The perspective view of Hengzhenbu Castle.

恒贞堡内的民居
Vernacular housing in Hengzhenbu Castle.

恒贞堡内的主街
The main street in Hengzhenbu Castle.

高家崖中的坡地建筑群
The architecture on the slope area in Gaojiaya.

城墙之下　Under the city wall.

4. 灵石市梁家堡

与红门堡方方正正的造型不同,灵石夏门镇的梁家堡依山就势而建,设计灵活,没有成规。

汾河流经夏门时西岸有一小段临河是峭壁的山岗,梁家堡便建于其上。为了让古堡顶端与汾河水仍有亲密接触,梁家在河边依峭壁直上直下地建了座40多米高的百尺楼,楼底层为安全只开一小门,楼梯段各层的开窗越高越多,越高越大,直至顶层变成敞廊,以将山河风光收进堡内,这座凌河矗立的宽大砖楼也成为梁家堡的标志。

古堡的正门在南面,经过两层楼的关帝庙进入头道堡门后拾阶而上,眼前景物步移景异。一会儿一座飞虹般的小桥掠过头顶,一会儿一座高墙横在眼前,而一转角又柳暗花明。这是座罕见的、地地道道的山地古堡,完全自然布局。可能是因为官气在梁家不稀罕,夏门梁家与静升王家不同,王家多富商,梁家多官,据说明清两朝梁家出了近200名大小官员,当然也需要梁家的官员们还留有读书人偏爱的闲情逸致,所以他们建设家园时不追求衙门状。

梁家堡百尺楼　The hundred-yard tower in Liangjiabu Castle.

4. Liangjiabu Castle in Lingshi city

Very different from the square shaped Hongmen Castle, Liangjiabu castle at Xiamen town in Lingshi city was built along the terrains of the mountains, improvise and flexible design is not restricted by rules.

While Fenhe River flowing through Xiamen, there is a short section of hills with sharp edge cliffs at the west bank, Liangjiabu Castle was built on that hilltop. In order to keep a close contact from the castle top to the Fenhe River, Liang family constructed a more than 40 meters-high tower straight up and down along the cliffs next to the river. For security, the ground floor only opened a small door, the opening numbers and size of windows for each staircase segment increases with the elevation level, the higher, the more and bigger, until the top level turns into the open corridor. This water-edge spacious brick building incorporates natural landscape and scenery inside the architecture; it has also become a symbol of the Liangjiabu Castle.

百尺楼内的楼梯
The staircases inside the hundred-yard tower.

沿山坡层层而上的堡内建筑
Buildings constructed along the elevated hill.

梁家堡周围的一些地名与隋末那段历史也有关系，如秦王岭、龙头岗、老生寨等。据说李世民撺掇他父亲在太原起兵后，在此与隋将宋老生发生过激战。然与张壁不同，没有记载与传说将这座古堡与隋末战事联系起来，梁家堡始建于明万历年间，与沁河古堡一样，与明末战事密切相关。

第三层堡门
The third layer of the castle gate.

山地院落的衔接
The joint section of courtyards at ground and higher levels.

窑洞式居住院落
The cave-style residential courtyards.

Climbing up the stairs from the castle's main entrance in the south, and passing by a two-story Guandi Temple, then entering the first gate to step on upwards, the landscape scenes change with every step. Never mind a rainbow-like bridge just flying pass the head, after a high wall blocking the view, then greet new vista at the turning corner. This is a rare and absolute mountain castle, completely natural layout. Maybe because Liang family did not concern much on showing off the official links, that is the difference between Xiamen Liang family and Jingsheng Wang family. Wang family has more wealthy merchants and Liang family possess more governing officials. It was said that Liang family produced nearly two hundred various ranking officials in the Ming and Qing dynasties. Liang officials also kept leisurely and carefree sentiments as the scholars prefer, so they built homes avoiding bureaucratic-like style.

Some of the place names around Liangjiabu Castle are linked to the history in Sui Dynasty, such as Qinwangling, Longtougang, Laoshengzhai and so on. It is said that after Li Shimin tricked his father to raise an army from Taiyuan, he had fierce fighting with Song Laosheng (a Sui commander) in Liangjiabu Castle region. Unlike the Zhangbi, no records and legends to link the ancient Liangjiabu Castle to the battles in late Sui Dynasty. Liangjiabu Castle was built during Wanli era in the Ming Dynasty, same as the Qinhe River castles, closely related to wars in the late Ming Dynasty.

堡内街巷
Streets and alley ways in the castle.

堡内的过街桥，连接上部两个院落
The over-street bridge connecting the two courtyards on top level.

五、陕、宁、鲁等省、自治区的民间古堡

1. 陕西横山县波罗堡

陕西的黄河支流无定河一线是汉与匈奴、唐与突厥、宋与西夏、明与蒙古长期对峙的前线，明长城就是沿无定河布局。从无定河至山西偏关，长城沿线留有许多至少是始建于明代的寨堡，一些寨堡清代、民国期间仍在使用，那时主要是为了防土匪、马贼之类，保留较完好的寨堡至今仍有村民居住。

陕北横山县境内的无定河边上现有响水堡、波罗堡等有名寨堡，它们建在河南岸的高坡上，易于瞭望又易守难攻。明代时寨堡里驻的应该多为军户，现在已是普通村民。

波罗堡的外围城墙高大，几座城门更为高大并有瓮城，临河一侧的城门口有座寺庙，寺内建筑依山势层层向上，顶上有座砖塔，为寺庙、也为古堡竖立了一个标志。堡内有低地、高台两片区域，两区内的房舍都破损严重，但因为有一条陡峭的、端头有座小神庙的蹬道连接两区，使堡内格局仍然比较清晰。

堡内最高处的房子最精致，古时它应是这里的首领住所，现在住着一户普通村民，房子是"土改"时分给他家的，一直种田的他们无力修缮，精雕细琢的一个窑洞四合院已经破败。

多重堡门
Multi-layer castle gates.

即将被黄土填平的瓮城
Barbican is about to be flattenly filled by the Loess.

Section 5 Vernacular Castles in Shaanxi, Ningxia and Shandong

1. Buoluobu Castle: Hengshan County, Shaanxi Province

From the Yellow River flows the Wuding River, along which was often the front lines of conflict between the Hans and the Huns, Tangs and Turks, the Song Dynastry and the Western Xia Dynasty, the Ming Dynasty and Mongolia. The Ming portion of the Great Wall lies along the Wuding River. From the Wuding River to Pianguan county, many Ming Dynasty fortresses were left along the Great Wall. Some fortresses were still in use during the Qing Dynasty and the Republic, mainly to counter bandits. Fortresses that still retain the majority of their integrity are still home to villages today.

Built on the southern high slopes along the Wuding River are Xiangshuibu, Buoluobu and other such fortresses, providing them with a clear view of the surrounding area and an easily defensible position. During the Ming Dynasty, fortresses were mostly military quarters, but have been long since become villages.

波罗堡的堡门，被现代车道割裂在高处，可见当初堡门前地势的陡峭
The castle gate of Boluobu, split by modern road. Obviously the gate was built on extreme steep terrain in the past.

城墙边的塔 High tower next to the city wall.

堡内窑洞 Caved living space within the castle.

117

董家堡大门　The main gate of Dong Family Castle.

2. 宁夏吴忠市董家堡

在宁夏的中卫市和吴忠市之间，有一座巨大的城堡形宅院，其主人是清代回族将官董福祥。董福祥幼年时家贫，没能读书，但善舞刀弄枪，适逢清末乱世，他在乡间聚众组织起一支小武装，占据一方自保，这种武装的性质多少都会有点土匪的成分，平日住在某座城堡里。后来小武装被清军招抚，董福祥于是成了官军军官。

庚子之变前，董福祥应诏入京。他支持义和团，率军协助义和团攻打北京东交民巷使馆区，但久攻不下。这时八国联军攻入北京，董福祥奉命保护慈禧西逃，因此护驾有功，当八国联军逼迫清廷严办主战和攻打使馆区的大臣时，慈禧无奈，又不想严办董福祥，便将董革职，永不叙用。

这时是1903年，而在此一年前，董福祥已在安排自己离职后的生活，他选中自己当年投身官军后第一次立下大战功的地方吴忠金积堡，兴建起一座大宅。在宅院外，董福祥又按西北地区城堡的形式为宅院加了一圈城墙，城墙外从前还有护城河、外寨。

Buoluobu Castle is surrounded by high walls guarded by several even taller city gates connected to barbicans. In close proximity to the northern gates, by the river, is a temple. The temple climbs up along the side of a mountain. At the top sits a brick tower that acts as a standard for both the temple and the fortress. Within the fortress lies a lowland and a highland that have both been severely damaged, a steep trail leading up from the lowlands to the highland where stand a small temple at the end of the trail, the layout of the fortress is clearly visible.

The houses highest up in the fortress are the most sophisticated. During ancient times, this used to be the lord's residence, but now houses an ordinary villager. It was given to his family during the "Land Reform." This family has been farming ever since and cannot afford repairs; a delicately crafted cavern courtyard has already fallen into ruin.

2. Dong Family Castle in Wuzhong City, Ningxia

Between Ningxia's Zhongwei City and Wuzhong City sits a huge castle-shaped house. Its owner is the Hui minority general of the Qing Dynasty, Dong Fuxiang. Dong Fuxiang's childhood family was poor, he wasn't able to go to school, but excelled at martial arts, just in time for the late Qing Dynasty chaos. He gathered a small army in the country to protect themselves and remain neutral, the nature of which will contain aspects of banditry, usually living in some castle or other. Later, a small armed Qing troop offered this small band amnestying enlistment, Dong Fuxiang later became a military officer of the Qing Dynasty. Before the crisis year of 1900 involving the Boxer uprising and the eight nation military invasion.

从城墙上俯瞰堡内院落
The perspective view of the courtyards from the city wall.

大门门洞　The main gate.

Dong Fuxiang was summoned by the emperor to Beijing. Supporting the Boxer Rebellion, he led his army to attack the embassy district on the eastern outskirts of Beijing with the Boxers, but too no avail after a drawn-out battle. Then, the Eight-Nation Alliance invaded Beijing, and Dong Fuxiang was ordered to protect the Empress Dowager Cixi to escape. When the Eight-Nation Alliance forced the Qing court to punish attackers of the embassy district, Cixi, helpless, not wanting to severely punish Dong Fuxiang, discharged him from service to the empire.

This occurred in 1903, but a year earlier, Dong Fuxiang already planned his life after the military. He chose the first place he won a battle after enlisting in the imperial arm, Wuzhong Jinji Castle, and newly built a mansion. Surrounding the mansion, he built a wall in the style of castles in the northwest, outside of which there used to be a moat and outer stockade.

院落建筑细腻的屋顶与简洁的城墙的对比
The building-style contrast between detailed roof and simple wall.

院落后部与城墙之间的环道
The ring road between the rear of the courtyard and the city wall.

魏家堡大门，门左侧和长方形城墙对角处各有一座圆形角楼
The main gate of Wei Family Castle. Two circular corner towers, one is placed at the left-side of the gate and the other is diagnostically at the rectangular corner of the castle wall respectively.

3. 山东惠民县魏家堡

魏家堡的前面有一个大水塘，当年挖这个水塘一来是为了使城堡有一个好风水，二来是为了取土将地基垫高，这是因为魏家堡所处的位置离黄河下游的大堤只有10公里，黄河在古代经常泛滥，将地基垫高可以减少洪水危害。城墙也是古代防御洪水的重要设施，所以魏家将住宅设计成古堡形式，应该也有对洪水的考虑。

魏家堡兴建于清末光绪年间，那时的山东开始出现号称"响马"的土匪，作为靠盐业专营、开钱庄当铺发迹的魏家，确实需要防范，近年在古堡里曾经出土了大量火铳用的铅弹。但从魏家是挖池塘而不是挖护城河这一点看，他们对古堡的美感、壮观似乎考虑得更多。建古堡的魏肇庆显然极爱面子，古堡前的两座旗杆象征着他的功名或官职，但那些都是他花钱捐的，不是读书考来的。

中国这些私人古堡的兴建对于社会来讲也算是一种固定资产投资，可以为一些乡民提供短期工作机会。据记载，在兴建古堡的1890年，黄河就曾经决口，惠民县灾民生活艰难，这时他们可以通过到城堡工地干活而获得一些钱粮。

魏家堡前的水塘
The water pond in front of the Wei Family Castle.

城墙上的马道
The bridle way on the city wall.

古堡内的院落　Courtyard within the castle.

3. Wei Family Castle in Huimin County, Shandong Province

The Wei Family Castle sits in front of a large pond was dug so that the castle can have good *feng shui*, and also to elevate the foundation. This was because the location of the Wei Family Castle lies ten kilometers downstream of the Yellow River enbankment. Since the Yellow River often flooded in ancient times, an elevated foundation reduces flood hazards. The walls were also in ancient times to protect against floods. Thus, the Wei Family Castle, although residential in nature, was built in the form of a castle due to the possibility of flooding.

The Wei Family Castle was built in the reign of the Qing emperor Guangxu, during that time, from Shandong Province came bandits known as "Xiangma". The Wei Family who prospered through selling salt, and opening banks and pawnshops, really needed to defend against. In recent years have, a large number of lead bullets used in blunderbusses have been discovered in the castle. However, since Wei Family dug a pond instead of a moat, they seemed to be value the aesthetic beauty of their castle than its defensive ability. Wei Zhaoqing, obviously very vain, placed two flagpoles in from of his castle to indicate his fame and status. However, he bought these with money and did not earn them by being clever and educated.

The construction of these private castles, for the society at large, can be considered a fixed asset investment, and can provide a short-term job opportunity for some of the villagers. According to the records, in 1890, the year the castle was built, the Yellow River burst. Huimin county victims lived hard lives, and could earn some money and grains by working at the construction site of the castle.

城门内部　　The inner side of the city gate.

4. 山东肥城市古堡群

肥城市位于泰山西面，春秋时期这一带是齐鲁边界，由于齐国筑有齐长城，而肥城周边的山上有许多石头的古城墙遗迹，所以有学者认为那是齐长城遗迹。

不过还有一个因素也可能会导致有古堡遗迹留在这片土地上，肥城这个地名源自这里曾经是肥族人的家园，肥族人是羌人的一支，"羌"字为"羊人"，即暗示羌人曾经长时间是游牧人。羌人的祖地在四川岷江等长江上游各支流的河谷中，他们从远古开始就逐步东迁，汉人的祖先很多就是东迁的羌人，如大禹就是一个东迁的羌人部落的首领。不仅现在的肥城人是羌人后裔，许多山西人，以及大量西北地区的人都是羌人后裔。

同样，汉魏时期的许多汉人也是羌人后裔，但在那时，他们的西面继续出现大量羌人。党项羌人是东迁羌人的一支，后来他们被隋朝降服，隋文帝曾经"谆谆教诲"党项羌人的首领："人生当定居"。他们定居了，不久建立了西夏王朝。

肥族人是什么时期定居在肥城的？那些古堡是肥族人还作为羌人时建的吗？或是后来当地人为避祸乱而建？我们无法得到准确答案，但可以肯定，肥族人刚迁来时与当地其他人群必有冲突，他们需要建城堡。羌人有建城堡的传统，现在他们的祖地那里的老民居还是以设防村寨结合碉楼的形式为主。

古堡城墙遗迹　　The remains of ancient castle wall.

4. Fei City Castle Group in Shandong Province

Fei City, located to the west of Mount Tai, during the Spring and Autumn Dynasty, was on the border of Qi and Lu kingdom. Due to the Qi Kingdom had built its own Great Wall, and the surrounding hills of Fei City contain many stone walls of ancient cities, some scholars believe that these are remnants of the Great Wall built by the Qi Kingdom.

There may be another reason why the ruins have remained in this land. The name "Fei City" is derived from the once the homeland of the Fei people, which are a part of the people of Qiang. The word "qiang" is made up of the words "sheep" and "people," implying that the Qiang people were once nomadic. The ancentral land of the Qiang people is along such rivers such as the Minjiang River, and other streams upriver of the Changjiang River, in Sichuan Province. The Qiang people steadily migrated east and became the ancestors of the Han people. Da Yu is a Qiang tribe leader. Not only are the Fei City people Qiang descendants, many people of Shanxi and the people of the northwestern territories are descendants of the Qiang.

Similarly, in Han and Wei dynasties, many Han people were Qiang descendants, but at that time, a large number of the Qiang people still lived in the west. Dangxiang Qiang people are one group of Qiang migrating eastward. They surrendered to the Sui Dynasty, Emperor Wen of Sui taught the leader of the Qiang: "Life should have roots." The Qiang settled down and soon established the Western Xia Dynasty.

When did the Fei tribe settle in Fei City? Were these castles constructed by the Fei tribe when they were still Qiang people? Or were later built by the locals to avoid misfortune and chaos? We can only guess, but there certainly was conflict between the Fei tribe and other local population, which would call for the construction of castles. The Qiang people also have a tradition of building castles, now, the old houses of their ancestral land are still fortified villages mainly consisting of Diaolou.

古堡内的房屋遗迹
The remains of houses in the castle.

第三章　闽粤赣 3 省的民间古堡

Chapter III　Vernacular Castles in the Three Provinces of Fujian, Guangdong and Jiangxi

闽粤赣古堡角楼上的枪炮口和雨水口
The firing hole and rain drainage hole in the Corner tower in castles in Fujian, Guangdong and Jiangxi provinces.

一、历史文化背景

1. 人口大迁徙

西北地区的羌人、胡人等向黄河中游迁徙可以被认为是中国历史中的第一次人口大迁徙。到西晋末年社会再次爆发危机时,民族冲突成为其中更重要的因素,早年内迁定居在黄河流域的游牧人这时纷纷建立自己的王朝,他们之间及与汉人的东晋王朝之间混战成一片,黄河流域的大量农耕人被迫向长江流域乃至更南方迁徙。随后,唐末、金灭北宋、元灭南宋、清灭明时期都出现过这种由北至南的人口大迁徙,由于这几次迁徙性质类似,可以将其合并视为第二次人口大迁徙。

由于远古时期中国的南方卑湿莽荒,黄河流域的人不愿意南迁,但随着人口增加,需要开垦新耕地,小规模的人口南迁实际上一直在进行,神农架、炎帝陵、舜帝陵等出现在南方就是一种体现。秦始皇统一中国后,继续向现在中国的南方沿海地区扩张,他派去的军队首领在秦朝灭亡时留在岭南自立为王,军队中许多人是北方人,他们曾经要求政府送一些北方的妇女过去,好让他们能娶妻生子,当然,他们也会与当地百越女子通婚。

在水稻种植技术日益成熟后,南方的粮食产量逐渐超过北方,自然条件也不断好转,这样,在北方有战乱时,就会有更多人南迁,在闽粤赣3省交界地区逐渐形成了一个庞大的客家人群,他们是最坚称自己的先祖是中原汉人、最坚守中原传统文化的中国南方人,而福建的闽人、闽粤交界处的潮人、广东的粤人虽然也称自己的先祖来自中原,但他们应该更多的是百越人的后裔。

广东省新丰县的围屋
A enclosure house in Xinfeng county in Guangdong.

广东省始兴县的碉楼
Diaolou in Shixing county in Guangdong.

迁徙人口往往会与土著居民发生争夺资源的冲突，闽粤赣地远山深，政府管理力度重要不足，所以不用等到各王朝末年，这些地方就会出现大量城堡。

2. 山海之间

除了北方人向闽粤赣地区迁徙，此地另一种间歇性的强大流动性人口来自海上或海边。

随着人口的迁徙和繁衍，闽粤赣地区也开始出现地少人多问题，沿海部分人口开始以海洋为生，除了捕鱼，也从事海上贸易。官方一向视民间海上贸易为走私，影响税收，也容易招引来海外危险人员，所以，历朝历代的政府都会间歇性地以打击海盗为名禁海，但禁海往往会使海盗行为大幅反弹，因为许多靠海吃饭的人一下子没了生路，只能彻底铤而走险。

沿海的混乱会使一些沿海居民内迁，一些海盗也会深入内地侵扰，这样都会使原来的内地居民感到危险性，他们纷纷建城堡自保，坚持住在沿海的居民肯定也会这样做。

碉楼内部的回马廊围绕幽深的天井
Horse-turning gallery way wrapped around the deep patio in the Diaolou.

碉楼大门，兼有装饰性和防御性，木门多包铁皮，门框上方有倒水的孔，防止攻击者烧门
The main gate of Diaolou has both decorative and defensive functions. Iron-sheet wraps wooden door, plus pouring water holes above the door to prevent the attackers letting fire on the door.

Section1 The Historical and Cultural Background

1. The Great Migration

Qiang and barbarian population in the northwest region migrating to the middle reaches of the Yellow River can be considered the first time Great Migration in the history of China. Ethnic conflicts become a more important factor of social crisis at the end of Western Jin Dynasty. The immigrated nomadic people during the early years have settled in the Yellow River basin then have gradually established their own dynasty. Lingering melees between them and the Han Chinese of the Eastern Jin Dynasty have been going on for a long time. A large number of farming people in the Yellow River Basin were forced migrating to the Yangtze River valley or even further south. Subsequently, these large scale of migration from north to south were common during the late Tan, Jin replacing North Song, Yuan taking over Southern Song and Qing destroying Ming Dynasty. As the migratory nature of these few times is similar, it can be regarded as the second time Great Migration.

In ancient times, China's southern regions were wet jungles with resources shortage, people based in the Yellow River Basin were reluctant to move south. But with the increasing population and demands on cultivation of new farmland, small-scale population moving southwards has actually been going on constantly. The existence of Shennongjia, Emperor Yan and Emperor Shun' Tombs in the south was a reflection of these migration. After the first Emperor Qin Shi Huang unified China, he continued to expand his control to China's southern coastal areas. The army chief he sent to the south in the ending period of Qin Dynasty has stayed in the Lingnan region and claimed as a king himself. Many northern people in the army had asked the government to send over some women from the north, so that they can get married and have children. Naturally, they would also have marriages with the local Baiyue woman.

As rice cultivation technology gradually matured, food production in the south increasingly exceeded the north, while natural conditions also continue to improve. Therefore, when there was war in the north, there would be more people migrated to the south. Then a large Hakka population gradually formed at the junction region of Fujian, Guangdong and Jiangxi provinces. These southerners group insist the most that their ancestors were the Central Plains Han Chinese, also firmly stick to the traditional culture of the Central Plains. Although Min people in Fujian Province, Chao people at the junction of Fujian, Guangdong, and Yue people in Guangdong also claimed that their ancestors originated from the Central Plains, but they should be mostly descendants of Baiyue people.

元朝用海军打击南宋时、明代中期以后的倭寇大爆发时、清代初期为了收复台湾实行"迁界禁海"政策时，都造成沿海居民大量内迁。明朝、清朝政府也在沿海修有大量城堡，其中的少数和少量民间城堡保留了下来。

3. 宗法与风水

山西那些建城堡的大家族都是因为富裕家族才大，没有富起来的家族一般都不大，但迁徙人群为了生存，就要首先集中家族力量，所以他们无论贫富都聚族而居，这样才能争夺生存资源。

要维持聚族而居，就要推行宗法制度，宗法制度反映在建筑上主要是讲究秩序和主次分明，大家族的大屋在建筑空间上也需要一定的秩序性。这种聚族而居的大屋体量必然大，同时在增强屋内空间凝聚力的要求下，这些大屋的内部布局都有较强的向心性，这已经为大屋设防提供了方便，而一旦专心设防，大屋很自然地就成为了一座城堡。

宗法制度带有相当强的强迫性，但光靠强迫效果一定不佳，还要给家族成员安全感、荣誉感和希望，才能更好地维护大家族。安全感由城堡因素解决；荣誉感首先是推崇祖先，他们都尊同姓的历史伟人为先祖；其次是修家谱、建

福建省厦门的沿海古堡，多为政府军队的炮台
Seaside castles in Xiamen, Fujian were mainly cannon platforms for governmental army.

福建泉港定楼，沿海民居少有紧邻大海的可能与取淡水不便有关
Dinglou in Quangan, Fujian. There was a little dwelling very near the sea, might relate to the inconveniency of getting drinking water.

Migratory population would often compete for resources and ignite conflicts with the indigenous peoples. In the deep mountains of Fujian, Guangdong and Jiangxi provinces, there were severely lacks of governmental controls, so no need to wait until the approaching ends of all dynasties, there would be a large number of castles in these regions.

2. Amongst mountains and oceans

In addition to the northerners migrated to Fujian, Guangdong and Jiangxi, and here another strong intermittent flow of population comes from the sea or the beach region.

As the population migration and reproduction, the problem of more people and less land occurred in the Fujian, Guangdong and Jiangxi joint region. The population in coastal areas started to rely on the ocean for living, in addition to fishing, they also engaged in maritime trade. Governments always regarded civil maritime trade as smuggling, it not only had bad impact on taxes, but also easy to attract overseas threats. Then the government of the dynasties would intermittently to impose activity bans on the sea in the name of combating piracy. However, the ban on the sea tend to make piracy rebounded sharply, because it suddenly cut off the lifelines of many people relying on the sea, left them out of choice but taking the risk of being pirates.

Coastal conflicts made some coastal residents moving further inland, and then some pirates also extended their intrusions in the mainland. These incidents made the original mainland residents feeling the threat and danger, so that they were urged to build castles to protect themselves, the same defensive reactions also came from the residents remained on coastal areas.

When Yuan naval army fighting against the Southern Song Dynasty, or during the Japanese pirates outbreak in the middle and late Ming Dynasty, or the implementation of the "coastal evacuation and ban on the sea" policy for recapturing Taiwan in the early Qing Dynasty, all had caused lots of coastal residents' migration. Governments in the Ming and Qing dynasties also built a large number of castles along the coast, minority of the official castles and a few private castles retained.

3. Patriarchal system/Patriarchy, and *Feng Shui*

In Shanxi, large families of those who built the castles were wealthy families, their wealth matches their big family scale, not rich families were generally not large. However migratory populations were different, they must first gather family strength in order to survive. No matter rich or poor they were all living together, so that being able to compete for the resources for survival.

To maintain living together, it is necessary to implement the patriarchal system, the patriarchal system reflects in the architecture is mainly pay attention to order and prioritize.

宗祠或祖堂中的祖先牌位
Ancestral tablets in ancestral halls.

宗祠，用青色留名来激励族人；希望则是用大屋及祖先坟墓的好风水来预示族中未来必有大富大贵之人光耀门庭。风水学是古代中国人搞营建活动时的主要指导理论，人们多信奉，只是闽粤赣人更为笃信。

二、赣南四角楼

笔者在前面说过，中国和欧洲都有四角楼那种古堡，它简洁经典的形式俨然是古堡建筑形式的一种原型，更复杂的古堡形式多是由它发展而来。

中国的古堡原型应该产生在黄河流域，但那里至今没有出土过四角楼明器，现存的古堡中也没有那样简洁的类型。那种明器都出土在南方，而且南方现存许多那种造型的四角楼实物，难道这种原型是在南方产生的？但南方人、特别是建了最多四角楼的客家人说这种形式是他们的祖先从中原带到南方的，是汉魏坞壁的形式。究竟如何，目前没有定论，不过，山西那种带看家楼的堡院与四角楼也非常神似。

闽粤赣3省都留存有大量四角楼，赣南、粤北的客家地区则留存着一批形式最经典的。

闽粤赣地区对古堡的称呼多为"围"或"楼"，只有少数称为"寨"或"堡"，但在古代，"寨"是那里比较常用的称谓，一般来讲，寨表示建筑比较简陋。

Large family house in the building space also needs to follow a certain order. This big house, the amount of living together will inevitably enhance the requirements of the inside space cohesion, the internal layout of these large estates have a strong concentric, which has a big house fortification provides convenient. Once a big house has focused on it fortification it naturally becomes a castle.

Patriarchal system maintains very restricted force and control, merely rely on the forcing effect this system could not work well. Additionally, it gives family members a sense of security, pride and hope In order to better maintain a large family. Sense of security addressed by the castle factors; sense of honour recognized by respected ancestors, they regarded the same surnamed great men in history as their ancestors; also compiled genealogy and built ancestral halls to inspire tribal members; hope indicated by remarkable *feng shui* of grand houses and ancestral graves to forecast the influencing offspring's future glorious achievements for the family. *Feng shui* is the main guiding theory for the ancient Chinese people engaged in construction activities, the rules most people believe, but people of Fujian, Guangdong and Jiangxi more devout.

Section 2 Sijiaolou in the South of Jiangxi

As the authors mentioned earlier, China and Europe all have the four-corner-tower style of castles. Its simple and classic form seems to be a prototype of the castle built form, more complex form of the castle has evolved from it.

The prototype of the castle should be appeared in the Yellow River Basin, but Sijiaolou form has not ever been discovered amongst the funerary objects, this simple type was not found in the existing castles either. This kind of funerary objects unearthed in the south, and many kinds of shapes exist in the south. Is it possible that this prototype is created in the South? But Southerners, in particular, the Hakka people built up many Sijiaolou said that this form was brought by their ancestors from the Central Plains to the south, in the form of Wubi in the Han and Wei dynasties. Not conclusive, however, the castle courtyard with watchtower in Shanxi and the Sijiaolou bore a striking resemblance.

Fujian, Guangdong and Jiangxi provinces have retained a large number of the Sijiaolou, Hakka areas in southern Jiangxi, northern Guangdong retained a number of the most classic form.

In Fujian, Guangdong and Jiangxi regions, mostly called the castle as the "Wei" or "Lou," only a few known as the "Zhai" or "Castle," but in ancient times, the "Zhai" was the more common appellation, in general, Zhai indicated the construction is relatively simple.

比较标准的四角楼　A standard four-corner-tower house.

1. 龙南县关西新围

由鄱阳湖溯赣江至今日江西省南部的赣州，再翻南岭至珠江两条主要支流北江、东江上游的路，是古代从北方下岭南的主要道路，今天江西省南部的龙南县、定南县至广东省东江流域的和平县、龙川县，以及北江流域的始兴县、翁源县等地，都是客家人聚居区，有大量的四角楼和四角楼变体存在。

龙南县城至关西镇的公路边不时有造型很标准的四角楼出现，著名的关西新围也是四角楼形式，只是它的体量太大了，称其为"围"确实贴切。所谓"围"可以指这座古堡建筑的整体，也可以指古堡的外围。外围主要有两种形式，一种是其自身包含房屋，防御性主要来自外墙高大、厚重；有时外墙内有穿廊，便于防守；另一种是实墙，防御空间在顶部。新围的外围属于第一种，但没有穿廊，外墙上基本不开窗。

新围的主人原来居住在新围旁边的老围里，老围是一座不规则形古堡，是当年大家族聚居的地方，后来族中有人靠贩卖木材致富，就在清代嘉庆年间自建了规则形的新围，围内院落极其讲究礼制。

关西新围内部巷道　　The roadway inside the Guanxi Xinwei.

1. Longnan Guanxi Xinwei

By the Poyang Lake travels upstream of Ganjiang River to Ganzhou in the south of Jiangxi Province, then climbs over the Nanling Mountain and follow the road arriving at the upper reaches of Beijiang and Dongjiang (the two main tributaries of Zhujiang), which is the ancient main route from the north travel down to the Lingnan. In present time, the areas from Longnan county, Dingnan county, in the south of Jiangxi Province to Heping county, Longchuan county at the Dongjiang River Basin in Guangdong Province, as well as Shixing county, Wengyuan county in the Beijiang River watershed, are Hakkas inhabited areas. There are many Sijiaolou and the variation of Sijiaolou exist.

Along the road from Longnan county to Guanxi town, the standard shaped Sijiaolou will appear here and there. The famous Guanxi Xinwei is also in the form of Sijiaolou, due to its volume is huge, referred to as "Wei" is indeed appropriate. The so-called "Wei" refers to the ancient castle building as a whole, can also refer to the periphery of the castle. The periphery structure mainly has two types. One type contains its own housing, the defensive function comes from the tall and thick external walls; sometimes it has open veranda gallery inside the external wall for easy defense. The other type uses solid wall, the

关西新围　The Guanxi Xinwei.

俯瞰乌石村
The perspective view of Wushi village.

关西老围　The Guanxi Laowei.

四角楼的一种变体
A variation form of Sijiaolou.

乌石围边的巷道
The roadway around the Wushi village.

2. 乌石村

乌石村位于龙南县最南部的杨村镇边，同样，那一带的公路边也有大量四角楼。

乌石村内也有几座四角楼，而村中最古老、最核心的建筑是巨大的乌石围，它的内部是一组方正的院落，外围呈不规则圆形，结合围前的半圆形水塘，大围整体形状仿佛一只乌龟，四面的角楼仿佛是乌龟的4只爪子，这种格局在风水学中是大吉大利的。

3. 燕翼围

位于杨村镇中心的燕翼围建于明末，在闽粤赣地区现存的古堡中是最古老的之一，这并不说明在更早的年代里南方没有古堡，只是因为南方气候潮湿，古建筑更不容易保留下来。

燕翼围主体呈长方形，它也有4个角楼，只是没有均匀地分布在长方形的4个角上，而是集中布局于两个角，但也可以从侧面保护4个边。围内的天井最初是空的，围子的房屋都在4边外围内，这就要求外墙足够厚，燕翼围的外墙厚度达1.5米，由外部0.5米厚的砖石墙和内部1米厚的夯土墙共同构成。天井中现有的房子是后来增建的。

燕翼围的大门
The main gate of Yanyiwei castle.

燕翼围的角部
A view of the corner of Yanyiwei castle.

defensible space is at the top within the external walls. The periphery of Xinwei is the first type, but it did not have the open veranda gallery, rarely has windows on the external walls. Xinwei's owner used to live in Laowei next to this new one. Laowei is an irregular-shaped castle is the space where a large family living together in old days. Later, members of the family got rich by selling timber, then self-built a regular shaped Xinwei in the era of Emperor Jiaqing in the Qing Dynasty. The inner layout of the courtyard adopted and focused on the ritual system.

2. Wushi Village

Wushi village is located at the most southern side of Yangcun town of Longnan county, along the road side in that area also has a lot of Sijiaolou.

Wushi village also has several Sijiaolou. The oldest, as well as the core building in the village is a huge Wushiwei, its interior is a collection of regular courtyards with irregular circular peripherals. Combined the periphery circular shape with the semi-circular pond at the front, the overall shape of Wushiwei like a tortoise, surrounded by corner towers as if the four claws of the tortoise, this pattern in the *Feng Shui* theory is good luck.

3. Yanyiwei Castle

Yanyiwei castle was built in the center of Yangcun town in the end of Ming Dynasty, it is one of the oldest one amongst existing castles in Fujian, Guangdong and Jiangxi regions. It does not mean that there is no castle in the south region in the earlier years, but the Southern climate is humid, ancient buildings are less likely to be retained.

Yanyiwei's main body is rectangular, it also has four towers, just not evenly distributed in the four corners of the rectangle, but rather focus in two corners, can also be protected the four edges from the side. The patio area is initially empty, houses are constructed inside the four-side peripheral, which requires external walls very thick. Yanyiwei external wall thickness is 1.5 m, constituted by external 0.5 m thick masonry walls and internal 1m thick rammed earth walls together. Existing house in the courtyard is a later additional build.

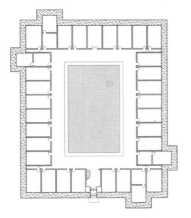

燕翼围平面图
The plan layout of Yanyiwei castle.

三、粤北四角楼

1. 和平县林寨四角楼

赣南和粤北交界的山区虽然分属于江西、广东两省,但在文化上都属于客家文化圈。与龙南县毗邻的广东和平县林寨镇也是一个四角楼集中的地方,其中石镇村本身是一座设防村落,外围城墙和城门仍然轮廓清晰,村内各家各户的房子多是中小型的四角楼或四角楼的变体,成片的四角楼散落在绿茵茵的稻田里,景象奇特而美丽。

相邻的兴井村则以两座大型四角楼为主,它们分别名为"谦光楼"和"颍川旧家",由于是建于清代末年的成熟型四角楼,所以它们的结构都非常紧凑,外围与内部院落式建筑是一个浑然的整体,而不是外墙包裹着一个院落,即它们是集中式建筑,而不是单元组合式建筑。以中轴线完全对称的布局显示礼制的严谨性,而环廊、穿廊、天井的组合又使内部空间显得非常丰富活跃,达到了很高的设计境界。

林寨的田野　　Surrounding field of Linzhai town.

Section 3 Northern Guangdong, Sijiaolou (Castle with four-corner-tower)

1. Heping County, Linzhai Town, Sijiaolou

The mountains region at the borderline of Southern Jiangxi and Northern Guangdong, although belonging to Jiangxi and Guangdong provinces separately, but culturally all belong to the Hakka cultural circle. Linzhai town in Heping county of Guangdong Province, adjacent to the Longnan county is also a Sijiaolou concentrated area. Shizhen village itself is a fortified village, the external walls and the gates are still clear outlined. The houses of each household in the village are mostly small and medium-sized Sijiaolou and its variants. Arrays of Sijaolou scattered in the verdant rice paddies, the scenery is spectacle and beautiful.

石镇村德兴居平面图
The plan of Dexingju, Shizhen village.

德兴居正面   The front view of Dexingju.

3 图:林寨四角楼的装饰细部
Three photos: the decoration details of the Sijiaolou in Linzhai.

The neighbouring Xingjing village featured by two large Sijiaolou, they were called the "Qianguanglou" and "Yingchuan Jiujia" respectively. Because it was built in a mature style of Sijiaolou in late Qing Dynasty, their structures are very compact. The external and internal courtyard style buildings are integrated as a whole, rather than the external walls wrapped around a courtyard. They are a centralized architecture Instead of a collection of unit modular buildings. Fully symmetrical layout along the central axis shows the rigor of the ritual system, and the porch, Loggia, patio combination of the internal space very abundant and active, reached a very high realm of design.

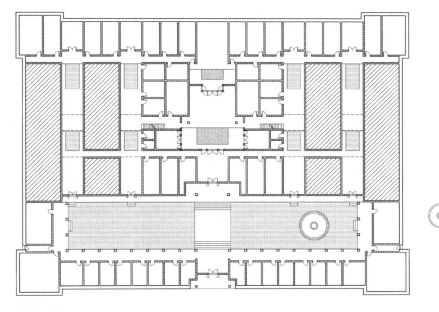

谦光楼平面图　The plan of Qianguanglou.

2. 始兴县的大碉楼

在关西新围后面的田心围那里,可以看到其外墙上有一座大碉楼,因其体量过大,不能视其为角楼或马面,而且其自身就带有马面,所以只能以碉楼视之。在赣南的寻乌县,这种碉楼更大,它们往往位于围屋之中或边缘处,但相对独立存在,相当于欧洲中世纪古堡中的中心塔楼。有时因其独立性更强,体量过大,它们自身就成为一座古堡。

从赣南如果沿梅关古道下岭南,过梅岭后是北江上游流域,那里有许多这种大碉楼。始兴县城东北面的东湖坪就有至少3座,其中的"永成保障"楼位于一座已经设防的大围屋后部,是堡中之堡。它的外形虽然不是典型的四角楼,但它显然是四角楼的一种变体,它更突出大楼的整体性,4个角部因各有马面衬托,仍然轮廓清晰。其内部为走马楼式布局,天井中有水井,如山西皇城相府的河山楼,在紧急情况时可以入内坚守,作为最后的避难所。

这种大碉楼的造型有着丰富的变化,有的只有两个角楼,有的就如四角楼明器所表现的,还有居中的望楼,有的没有角楼,只是将一个四合院升高了,但也是古堡的意象。

寻乌县的大碉楼
The Large Diaolou in Xunwu county.

关西田心围的碉楼
The Diaolou of Tianxinwei, Guanxi.

2. Diaolou in Shixing County

Tianxiwei is located behind the Guanxi Xinwei, you can see a large Diaolou on the outer wall. Because of its excessive body, it can not categorised as a turret or corner tower, but viewed as a main tower containing its own turret instead. In Xunwu county in south of Jiangxi, there are more of such main towers, often located inside the enclosure houses or at the edges, but relatively kept their independent existence, which is equivalent to the Tower in the medieval castle in Europe. Sometimes because of their greater independence, and excessive body, turning themselves into castles.

From southern Jiangxi along the Meiguan Road travelling down to Lingnan, over the Mountain Meiling you will reach the upstream basin of Beijiang River, where many of these Diaolou exist. At least three towers can be found in Donghuping in the east of Shixing county, amongst them *"the Yongcheng Baozhang"* tower is located at the rear of the large forted enclosed houses, which is the castle in Castle. Its shape is not a typical Sijiaolou, but it is clearly a variant of the Sijiaolou. Four corners are showed off by turrets on each one, highlights the integrity of the building with a clear outline. With its internal circular gallery layout, there are wells in the courtyard, such as the of Huangcheng Xiangfu in Shanxi, in case of emergency, it can be guarded securely and used as the last refuge.

The shape of these kinds of Diaolou has a wealth of changes, some only two turrets, some have a central tower, as shown in the Sijiaolou funerary objects, and some do not have turrets, just like an elevated courtyard, also an imagery of castles.

体量较小的独立式碉楼
Small-scale independent Diaolou

"永成保障"楼正面,下层只有枪炮孔,上层的窗户也很小
The front view of "the Yongcheng Baozhang" tower. Only firing holes in the lower floor level, windows on the higher floor levels are also very small.

"永成保障"楼位于围屋后部
The Yongcheng Baozhang is located at the rear part of the round house.

楼内天井　The lightwell in the tower

台兴县罗坝镇的长围也位于一座围屋后部
The Changwei is located behind another round house, Luobai town, Shixing county.

长围的楼门
The gate house of Changwei.

有外围护的四角楼，角楼只高起，不外凸
The Sijiaolou with periphery, the turret is high up but not bulging outwards.

由四合院升高形成的碉楼，没有角楼
The Diaolou formed by elevated courtyard without turret.

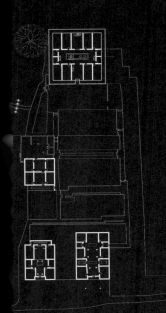

"永成保障"楼与围屋平面图
The plan of "the *Yongcheng Baozhang*" tower and the enclosed house.

只有只面的碉楼 Diaolou with turret only

3. 司前镇四角楼

始兴县南部司前镇里的燕翼围是一座比较经典的四角楼，它主体的平面是正方形，四角有正方形的角楼，角楼的宽度是主楼墙面宽度的1/2，加之楼高适度，使楼的外表比例匀称，极具体积感。

楼内楼梯在门厅两侧，起跑段很宽，上至一定高度后，由于底层1米多厚的外墙变薄，楼梯便利用节省出的空间沿墙体上行。天井内只在3层设木挑廊，使空间不致逼塞。大门一面和两个侧面的房屋进深相同，大门对面的房屋进深要大一些，因为有祖堂，敬祖是宗法制度的中心之一，闽粤赣老民居中最重要的房屋都是祖堂，它一般的位置都在大围、大楼的中轴线尽端。黄河流域的古堡中虽然也有祖堂或祠堂，但位置不如南方显赫。

总之，各种细致入微的设计使这座燕翼围可以作为四角楼的一个样板。

司前镇燕翼围的正面　　The front view of Yanyiwei, Siqian town.

3. Siqian Town Sijiaolou

The Yanyiwei in the south of Siqian town, Shixing county, is a very classic Sijiaolou, wihich has four-corner tower. It has a main square plane, with square corner tower in the four corners. The width of the corner tower is 1/2 of the width of the main building wall, coupled with moderate storey height, the Sijiaolou appears well-proportioned with great sense of volume.

Within the building, stairs are on both sides of the hall, the starting segment is very wide, up to a certain height, due to the thinning of the underlying more than 1 meter thick external wall, stairs extended upwards in the widened space along the wall. Wooden veranda gallery is only placed on the third storey in the patio, so that space will not have much blockage feel. Houses at the main entrance side and the two other sides have the same depth, houses opposite to the main entrance have larger depth, because there are ancestral hall on that side. Ancestral worship is one of the core aspects in the patriarchal system. The most important room in old residential houses of Fujian, Guangdong and Jiangxi is the ancestral hall, it is normally located at the far end on the central axis of the great enclosed houses or Keep Towers. Although there are also ancestral hall or shrine in the castles of the Yellow River Basin, but the location in not as prominent as in the southern castles.

In summary, the meticulous design of this Yanyiwei can be ranked as a model of Sijiaolou.

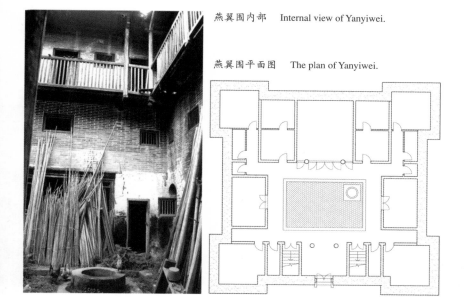

燕翼围内部　Internal view of Yanyiwei.

燕翼围平面图　The plan of Yanyiwei.

4. 隘子镇满堂围

始兴县隘子镇位于司前镇西面,镇外由3座大围屋组成的古堡群被称为"满堂围",建围的家族也是靠木材贸易致富,于清代道光至咸丰年间,用了近30年的时间建成这组建筑。

3座围中,居中的"中心围"的中心是一座大型四角楼,其中心进一步包含着一座四合院式的、带望楼的碉楼,四角楼外还有一圈比较低矮的外围,外围后部还有两座角楼。

左侧的"上新围"也有4座角楼,只是整体较低矮,围屋面积又大,使其在体积的比例上更显低矮,弱化了四角楼的轮廓。

从后部俯瞰满堂围　The perspective rear view of Mantangwei.

满堂围正面　　The front view of Mantangwei.

4. Mantangwei in Aizi Town

Aizi town is located in the west of Siqian town in Shixing county, the castle group of three enclosure houses outside the town is called "Mantangwei", The family relied on the timber trade and got rich, in the Qing Dynasty Daoguang to Xianfeng years, they used nearly 30 years to complete the construction of this group of buildings.

Amongst the three Weis, the center of the central Zhongxiwei is a large Sijiaolou, the center further contains a courtyard Diaolou with the watchtower. Outside the Sijiaolou, there is a lower periphery enclosure house, at the rear of the periphery there are two corner towers.

The left side, the "Shangxiwei" also has four corner towers, but an overall not very tall shape plus very large area, makes the area looking even more lower, weakening the contours of the Sijiaolou.

中心围内部　　The inter part of the central Wei.

5. 翁源县的多角楼

在始兴县南面的翁源县境内,规则的四角楼不多见,但有许多"多角楼",即平面为不规则多边形的围楼,其转角处多有角楼。这种形式在作为原型的四角楼出现前应该已经存在,它容易自然而然地生成,但如果是设计灵活而考究,虽然体型不规则但造型规整的多角楼,就应该是四角楼的变体了,是四角楼这种原型的变形。

江尾镇的湖心坝村是个由众多中小型古堡组成的大围村,那些古堡包括一座不规则椭圆形的围楼、几座围屋和十几座多角楼,不规则椭圆形的"长安围"是村中沈氏家族在此地建的最早的围屋,始建于明代中期,现有建筑重建于清代乾陵年间。现存的多角楼也是清代的构造,多角楼自身的造型更有古堡建筑特殊的美,而多角楼之间的巷道因为有分布灵活的角楼,从而产生出独特的街景。

水口镇的多角楼 Multi-corner tower in Shuikou town.

湖心坝的街景 The street view of Huxinba.

角楼上的马面 The turret of the corner tower.

5. Wengyuan County's Multi-corner towers

Within the boundary of Wengyuan county in the south of Shixing county, the regular shaped four-corner buildings are rare, but many of the "multi-corner towers" that the plane is irregular polygons, mostly with towers at the corners. This form should already exist before the prototype of the Sijiaolou, it is naturally designed to be flexible and elegant. The irregular body forms the regular shape of multi-corner towers, it should be variants of the Sijiaolou.

In Jiangwei town, Huxinba is a village composed by many small and medium-sized castles, The castles include an irregular oval enclosed house, several rectangular and more than a dozen multi-corner towers. The irregular oval "Chang'anwei" is the earliest enclosure house in the village built by Shen's family. It was built during the mid of Ming Dynasty, and rebuilt during Qianlong era in the Qing Dynasty. Existing multi-corner towers are also the structure of the Qing Dynasty, multi-corner tower shape expressed the special beauty of ancient castle construction, and the roadway between the flexible distributed corner-towers created a unique street view.

湖心坝的两座多角楼　　The two multi-corner towers in Huxinba.

四、沿海古堡

与山西沁河古堡集中兴建于明末陕西饥民攻掠时期的情况类似，现存的福建沿海古堡中最早的一批集中兴建于明代倭寇肆虐时期。明代万历年间的《漳州府志》中有这样的记载："嘉靖四十等年以来，各处盗贼生发，民间团筑土围、土楼日众，沿海地方尤多。"

又据明代的《漳浦县志》记载："方倭奴初至时，挟浙直之余威，恣焚戮之荼毒，于时村落楼寨望风委弃，而埔尾独以蕞尔之土堡抗方张之丑虏，贼虽屯聚近郊，迭攻累日，竟不能下而去。自是而后民乃知城堡之足恃，凡数十家聚为一堡，砦垒相望，雉堞相连。"

这种情况与窦庄古堡对沁河人的示范效应一样。

Section 4 The Coastal Castle

In Shanxi Qinhe castles were intensively built during Shaanxi starving riot period in the late Ming Dynasty. Similarly, the first batch of existing coastal castles in Fujian was built during Japanese pirates ravaging period in the Ming Dynasty.

In *Zhangzhou Prefecture Records* of the Ming Dynasty in Wanli era, the record stated: "During the Jiajing's four decades, thieves exist throughout the civil society, more and more civilian people built fortified villages together, particularly in coastal areas".

Also, the Ming Dynasty *Zhangpu County Record* recorded when the pirates arrived in the region, forced on with the remaining power from Zhejiang area and carried on rulelessly killing and looting the locals, people have to give up their villages and fleet. Only the people in Puwei used a small castle to fight the invasion and defend their homes. Pirates gathered outside the castle and attacked several days but had not been able to capture the castle, consequently withdrawal. Since then, people knew the fact that the castle could be against an enemy entrenched, then dozens of families gathered to build a castle together, the neighbouring castles overlook and interconnect to each others.

This situation was the same as Douzhuang Castle's demonstration effect on Qinhe people.

1. 闽东南的寨中楼

清代康熙年间,为了孤立和收复台湾,在东南沿海实行过"迁界禁海"政策,沿海 10～50 公里的地域内,民众全部内迁,民房全部拆毁,包括民间城堡。所以,那里清初以前的古堡几乎无存,如埔尾"蕞尔之土堡"肯定要被强拆,只有少数大型的古堡因为清军要利用才幸存下来,它们多建于倭寇第一次大规模袭击福建沿海以后,大约是 1560 年代,漳州市漳浦县城东面 30 公里处的赵家堡虽然建造年代略晚,但也属于这种类型。

建造赵家堡的赵氏家族据称是宋代皇室的后裔,堡内最有古堡形态的建筑"完璧楼"建于明代万历二十八年(1601 年),其形式与粤北那种由四合院升高形成的古堡一样,作为家族和乡里在紧急情况下的避难所。由于倭寇之害持续不断,赵家又在 20 年后扩建了高 6 米、阔 2 米、周长 1 公里的外围城墙,加建城楼,形成现在的样子。

完璧楼　Wanbilou tower.

1683年台湾战事结束，清朝政府逐步"复界"，这时虽然已经没有大的战乱了，但民间聚族械斗之风又起，这是迁徙地区常有的现象。一些大家族为了安全，也为了展示实力，在复界区兴建起新的城堡，泉州市泉港区前黄镇黄氏家族的"定楼"就属于这种类型。

　　建于清代乾隆年间的"定楼"建筑群的模式与赵家堡类似，只是建筑的布局、质量更考究，一座规整的、有外围的方形院落中间，是一座壮观的大石楼，其建筑材料几乎全部是巨大的石块，包括楼板。由于大石楼基本处在院落正房的位置，这种格局又使人想起山西沁河有碉楼的堡院格局。

　　"定楼"无疑是现存非常出色的一座沿海民间古堡，但遗憾的是，我们没能找到那种真正面朝大海的民间古堡，如惠安县的崇武古城等都是当时政府的军事卫所。

福建惠安崇武古城，明代的军事卫所
Chongwu city, Hui'an County, Fujian, a military castle in Ming Dynasty.

1. Fujian Southeast Keep Tower

During Kangxi era in Qing Dynasty, in order to isolate and recover Taiwan, the "Evacuation and ban on fishing" policy was implemented in the southeast coast. Within a geographic area of 10 to 50 km along the coast, people all moved inland, houses including vernacular castles were all demolished. Almost none castle before early Qing Dynasty remained, the smallest castle in Puwei sure to be demolished. Only a few large castle survived because the Qing's army kept these for their own use. These castles mostly built after the first large-scale pirates' attack on the coast of Fujian, which is about the 1560's. Zhaojiabu in Zhangzhou City, 30 kilometers east of Zhangpu town, although was constructed a little late, but also falls into this category.

Zhao family, who constructed the Zhaojiabu, said to be descendants of the royal/imperial family of the Song Dynasty. The most castle shaped building Wanbilou was built in the Ming Wanli 28 (year 1601). It has the same high formation as the elevated courtyard in the northern part of Guangdong, and used as refuge for the family and villagers in case of emergency. Due to the continuous threats of Japanese pirates, Zhao family made expansion to the original castle after 20 years. Additional external walls of the 6 meters high, 2 meters wide, the circumference of 1 km were constructed to form the current castle.

In 1683 end of the Taiwan war, the Qing government gradually "Restore the boundaries", then there was no major war, but the tribal disputes and fighting underway, which is the common phenomenon of migration areas. Grand families, in consideration for safety as well as showing off their wealthy, started the construction of new castles in the complex boundary region. Huang family's Dinglou, in Qianhuang town in the Quangang district of Quanzhou city, falls into this category.

The building group of "Dinglou" built during the Qianlong years, has similar mode as Zhaojiaobu, only the layout of the building and the quality are more sophisticate and elegant. A spectacular stone Keep Tower is located in the middle of a square courtyard with regular periphery. All most all the building materials are huge stones, including the floors. The Keep Tower is placed at the location of the main house in the courtyard, this pattern is reminiscent of the castle with tower pattern of Qinhe area Shanxi.

"Dinglou" is undoubtedly the excellent one of existing coastal vernacular castles. Unfortunately, we could not find the type of vernacular castles truly facing the sea, such as the Chongwu ancient city in Hui'an county, which were government's military defending compounds at the time.

定楼正面　The front view of Dinglou.

充当院落中正房的定楼
Dinglou used as the main house in the centre of the courtyard.

定楼内部　The inside of Dinglou.

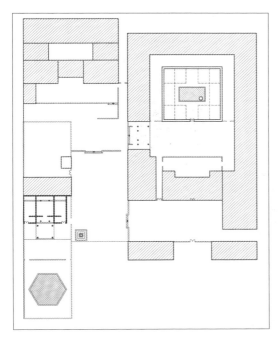

定楼组团平面图
The plan of Dinglou group.

159

2. 潮州堡寨

广东省潮州市与闽南漳州紧邻，历史大事记与闽南类似，潮州人与闽南人也都属"福佬人"，但潮州的文化也有自己的特点，在民间古堡建筑方面，除了圆土楼（后节详述），主要是圆寨和方寨，无论方圆，这些堡寨均尽量临水而建，利用水面增强防御性，同时方圆只是外形，寨堡内部的建筑群落均以方格形组团为单元。

民间堡寨的外围多是高大、外墙加厚的房屋，不仅仅是道实墙体，而当地同时代的政府军事卫所，外围多是一道墙。

潮安县的象埔寨和龙湖寨最古老、最著名，但它们都大得像一座城镇。邻近象埔寨的长美寨规模适度，呈现着一副中小型方形古堡的样子。寨门直通主街，主街尽端是一座小神庙，两侧是整齐排列的方形院落，这种格局与山西晋中的晋商古堡相仿。

长美寨周边的水面大部被填了，揭阳市大溪镇井美村由于地处偏僻，环村水面还大部幸存。揭阳古属潮州，虽然有大量客家人口，但文化总体上偏向潮州，井美村中的李氏家族也是潮州人，堡村的格局与长美寨不同，其主轴线不是主街，而是主要宗祠的中轴线，外围形状结合了水体的形状。

长美寨内的主街
The main street in Changmei village.

2.Chaozhou Castle

Chaozhou city in Guangdong Province close to Zhangzhou in southern Fujian, has similar historical memorabilia as southern Fujian. Chaozhou and southern Fujian people are considered Hoklo, Chaozhou's culture has its own characteristics, in addition to the round earth buildings (which will be detailed in following sections), mainly round castle and square castle styles as the vernacular castles. Regardless of the square or circular shape, these castle were built as near the water as possible, using the water to enhance the defensive function. Either square or circular was just the external shape, square grid unit to form the architectural community within the castles.

The periphery of vernacular fortresses is mainly tall houses with thickened external wall, not just a solid external wall. However, the peripheral of the governmental military guard buildings is mostly only a wall.

Xiangpu and Longhu villages in Chao'an county, are the oldest and most famous castles, but they are as big as a town. Near Xiangpu village, Changmei village has modest scale like a small and medium-sized square castle. Village gate straight through main street, a

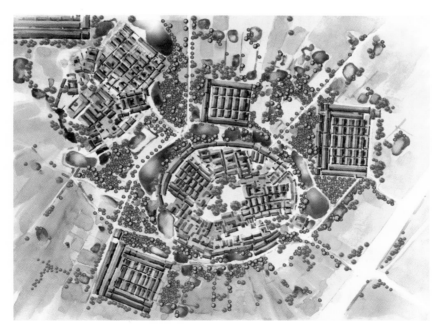

依照卫星地图绘制的潮州村寨结构图
The layout of villages in Chaozhou based on satellite images.

161

井美村外围　　The external perimeter of Jingmei village.

small temple placed at the end of main street, flanked by neatly arranged square courtyards, this pattern is similar to the Shanxi merchants' castles in the middle of Shanxi.

Most of the water body surrounding Changmei village has been filled, due to the remoteness of Jingmei village in Daxi town of Jieyang City, most of the water body wrapping around the village survived. Jieyang bellows to Chaozhou in ancient times, although there were a large number of Hakka population, but the culture in general tend to bias Chaozhou. Li's family in Jingmei village is also originated from Chaozhou, pattern of its fortified village is different from Changmei village. Its main axis is not the main street, but follows the central axis of the main ancestral hall, the peripheral shape is in combination of the shape of the water body.

第四章　闽南粤东的圆土楼

Chapter IV Round Earth Buildings in South of Fujian and East of Guangdong

福建永定县洪坑村振成楼鸟瞰图
The bird's eye view of Zhenchenglou in Hongkeng village, Yongding county, Fujian Province.

一、对圆土楼来源的探讨

中国福建土楼是独一无二的民居形式，也是一种独特的古堡，特别是其中的圆土楼，以其特立独行的形态引发世人好奇，人们最好奇的问题是：圆土楼是什么人、什么时候、因为什么而设计出来的？

1. 对现有解释的疑问

目前，对此问题的解释是，由于漳州的史料中最早对土楼的记录起自明代嘉靖倭乱时期，而且现存最古老的圆土楼多是漳州与明代抗倭有关的土堡、土楼，所以圆土楼应该是由漳州抗倭时期的圆形军事寨堡逐步演变来的。

然而，这样无法解释如下问题：

首先，史料中记载在倭乱之前，漳州土楼、土堡是"旧时尚少"，不是没有，而"旧时"到底旧到什么时候？同时，史料中也有记载，早在唐宋时期，这一带的畲民就用寨堡抵制官军。

第二，漳州没有更古老的圆土楼，不说明其他地方也没有，潮州饶平县有些圆土楼据称始建于明代初年。

第三，如果圆土楼是由圆形寨堡演变来的，那么，为什么其他也有大量圆形寨堡的地方没有演化出其他圆形建筑，只有闽粤这一小片地区产生了独特的结果。

第四，如果圆土楼是由圆形寨堡演变来的，那么，当圆土楼还没有成为一种建筑风格时，建圆土楼的目的应该都是追求防御性。但我们在闽南和潮州都可以看到许多没有合拢的弧形土楼，有的可能是因为一直没完工；但有的因为其弧形的半径过大，根本无法合拢；有的完全可以合拢，但楼又沿着别的方向续建了很多，但就是不合拢，也就是说，其建造者根本就没有追求防御性。那么，他们为什么建弧形的楼？而且，有些弧形楼没有外围土墙，这可以说明，这种弧形楼与圆形寨堡没有关系，但它们应该与圆土楼有直接关系。

第五，历史当然有特殊性、偶然性，但如果有逻辑性，就会更令人信服。而且，历史演变一般地都有深层逻辑性。

Section 1 Investigate/Trace the Source of the Round Earth Buildings

Earth building in Fujian, China, is a unique form of residential structure, and also a unique castle, especially the round earth buildings, it's maverick form leads to the world of curiosity. The most curious question is: who, when and what the designs for?

1. Doubt on the existing interpretations

At present, the interpretation for this question is: based on Zhangzhou's historical data, the earliest record of earth buildings dated back from Japanese piracy period in Jiajing era in the Ming Dynasty, and the oldest surviving round earth buildings are mostly the earth forts and earth buildings for against pirates in Zhangzhou region. Therefore, the round earth buildings should be a gradual evolution of the circular military fortress during the fighting against Japanese pirates period in Zhangzhou area.

However, this interpretation can not explain the following questions:

A. The historical data recorded before the Japanese piracy chaos time, Zhangzhou's earth castle and earth building were rare, also the "old" indicated when is the exact time not clear. Meanwhile, in the same historical records, as early as the Tang and Song dynasties, the ethnic She people already used Fortress to resist government troops in this area.

B. No old round earth buildings in Zhangzhou, does not mean no existence elsewhere. Some round earth buildings found in Raoping county in Chaozhou were allegedly built in the early Ming Dynasty.

C. If the round earth building is the evolution of circular fortress, why no other circular-shape buildings evolved from a large number of round fortress in other areas, it has the unique transformation only in the small areas in Fujian and Guangdong.

D. If the round earth building is the evolution of circular fortress, then, when the round earth building has not yet become an architectural style, the purpose of building the round earth buildings should all be pursuing the defensive function. However, in southern Fujian and Chaozhou, we can see many non-enclosed arc earth buildings. Some may have not been completed; but some because of its curved radius is too large, do not close; and some can close but the continued construction is along the other direction, surely not seeking closure. It is very obvious that its builders do not pursue defensive function at all. So, why they built the curved buildings? Moreover, some curved buildings has no external earth walls, which could explain this type of arc building has no links to round Fortress, but they should have a direct relationship with the round earth buildings.

福建南靖县的翠林楼,号称最小的土楼,位于一组民居中间,仿佛是一座古堡中的大碉楼
The Cuilinlou in Nanjing county, Fujian. It is the smallest earth building amongst group of houses, as if a big Diaolou in a castle.

2. 圆形的神圣性与原始性

军事寨堡出现圆形不令人奇怪,因为它们的建造都比较匆忙,在有好的防御性的前提下,多快好省是最重要的。在要保护的地盘面积相等的情况下,圆形的边长要比方形至少短10%,而且圆形有更强的土地条件适应性。

圆形寨堡对于施工也不会增加难度,因为,不论是垒石头还是夯土,圆形墙体都不会比直线墙体麻烦。但建圆形土楼就不同了,圆楼的木结构要变得麻烦很多。那么,人们建圆土楼的原因应该只有两点:一是喜欢圆形,并且有钱有闲,这应该是晚期成熟型土楼的主要成因;二是文化传统和民间习惯使然,其中文化传统是最关键的因素。

圆形因为其有自然原始性,在中国的最远古时期曾经非常流行,如东北地区的红山文化和黄河中游的半坡文化遗址中,建筑多是圆形的。但最迟到西周时期,中国北方的人居建筑就不用圆形了,圆形基本上成为天、神灵专用的形式,所谓"天圆地方",人居在地,从此直线成为人居建筑主导。

E. The history of course has its special, incidental progress, but if there is also a logical evolution, would be more convincing. Moreover, the historical evolution generally has a deeper embedded logic.

2. The sanctity and the primitives of round shape
It is not surprising that military fortress has round shape, because their construction is more rushed, under the premise of a good defense, faster and more economical is the most important. To protect an equal site area, the perimeter of circular/round shape is at least 10% shorter than a square, and the round has stronger land adaptability.
Round Fortress will not increase the difficulty for the construction, because, regardless of base stone or rammed earth, a circular wall is not more troublesome than a straight linear wall. But building a round earth building, the construction of the wooden structure becomes more complicated and challenging. Reasons for people to build the round earth buildings should be only two points: firstly, like round shape, and also had money to spare. This should be the main causes of the late mature earth buildings; Secondly, cultural traditions and folk customs are the driving forces, in which cultural traditions is the most critical factor.
Because of its natural original, Round once was very popular in China in the most ancient times, such as in the Hongshan culture remains of the Northeast and Banpo cultural heritage sites in middle reaches of the Yellow River, the buildings are mostly circular. But as late as the Western Zhou Dynasty in northern China's habitat building not round, round basically become the special form for the gods and spirits, the so-called "Sky round and earth square", people's habitat to the ground, since then straight lines led the form for dwelling structures.

湖心坝中的长安围　　Chang'anwei in Huxinba.

西方也有类似的观念，如西班牙巴塞罗那的著名建筑师高迪说过："曲线属于上帝，直线属于人类。"不过，他强调的是艺术追求问题。

在礼制观念产生后，中国古代人居建筑、特别是民居中，就更难见圆形因素了，几乎只有圆土楼、客家围龙屋、闽南潮州弧段楼、蒙古包等例外。一般的圆形寨堡不能算圆形建筑，因为它们只有城墙是圆形的，设计它们不需要有设计真正圆形建筑的意识。而闽粤赣内地、潮州沿海还有不少由弧形的外围房屋围成的围屋、围村、圆寨，如前面介绍过的广东翁源湖心坝中的"长安围"及其邻近的"思茅墩"（两围始建年代均早于倭乱时期），但如果说是这种圆围演化出了圆土楼，就如说圆形军事寨堡演化出圆土楼一样，缺乏必然的逻辑性，但它们毕竟比圆形军事寨堡更接近圆土楼。

3. 关注井

如果说四角楼是古堡的一种原型，那么四合院就是中国民居的一种最重要的原型，这个原型背后的逻辑性最为清晰，那就是井田制。

凿井技术在中国据传由神农发明，人类学会凿井取水是一件意义重大的事，因为从此人类耕种和生活就不必过分依赖河流了，这样不仅安全多了，可居住的范围也大多了，中国农耕人的聚落都围绕着井布局。

南方院落中的天井
The patio of the courtyard in the south of China.

中文中的"井"字在最古老的甲骨文和金文中就基本是现在这个架构，两横两竖表示井口的木架，在篆字中，其中心还有一个圆点。这种字形的图式作用后来被用来划分土地，进一步形成一种居住模式、社会制度、社会秩序、文化符号等，一块中间有井的方形土地被"井"形划分为9块，居中有井的一块为公田，周边的8份平等地分给8户人家，8户人家除了耕种自己的土地，还要共同耕种公田，公田的产出为公共储备。这种公平、有协作精神、有秩序（井井有条）的结构得到周代统治者的青睐，井成为一个土地单位，也是一种社会单位，政府据此计算税收。

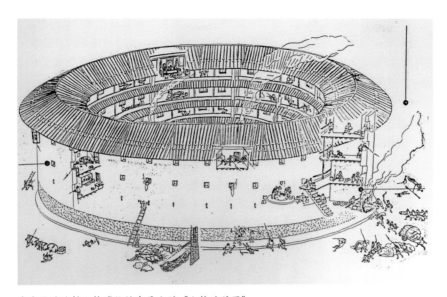

永定县洪坑村土楼博物馆中展出的《土楼攻防图》
The illustration of *the earth building's defense plan*, at the exhibition of earth buildings museum in Hongkeng village, Yongding county.

In the West there are similar concepts, such as the famous Spanish architect Gaudi from Barcelona said: "Curve belongs to God, a straight line belonging to the human." However, he stressed the artistic aspect of architecture.

After the formation of the concept of ritual, the ancient Chinese dwellings, especially in residential areas, the circular factor is even rarer. Almost only the round earth buildings, Hakka Weilong House, southern Fujian and Chaozhou arc Building, Mongolian yurts and other exceptions. Round fortress cannot be considered rotunda, because they only have the circular designed walls, no need to be aware of the critical design principles and details as for rotunda. In the inland areas of Fujian, Guangdong and Jiangxi, and the coastal areas of Chaozhou, there are lots of round houses surrounded by the arc of the periphery of housing, walled villages, and round village, such as Huxinba village in Wenyuan county, Guangdong, which we described earlier, "Chang'anwei" and its adjacent "Simaodun" (the two Weis were both built before the Japanese piracy chaotic period). If assume round earth buildings were evolved from this circular circumference structures, that round earth buildings would be evolved from circular military Fortress, apparently, it is lack of the obvious logic. After all, if compared to a circular military Fortress, they are more close to the round earth buildings.

井田制在春秋末期逐渐瓦解，商鞅变法的重要一条就是废井田而开阡陌，以适应更大规模的农业生产。但在春秋战国的乱世中，井田被描述的井然有序的社会图景成为以孔子为代表的立志结束乱世的一群文化志士和理想主义者试图恢复的乌托邦。但毕竟，井田模式比较适合氏族、部落规模的人群采用，随着社会组织规模的不断扩大，井田这种生产关系就越来越妨碍生产力发展，但在严重的社会腐朽混乱中，人们有今不如昔的感觉，会对过去过度怀念。

井田制在中国文化中刻下了深深的烙印，中国人住在城镇是住在"市井"中，离开家园是"背井离乡"；古代村镇的模式多是井田的意象，井作为"共汲之所"通常位于村镇的中央，围绕井会出现一个广场，人口居住在广场四周；井有时也会在村口广场上，但总之，因井总会形成一处公共空间，该公共空间一定会形成村镇公共意志的场所，从而成为村镇的代表符号。

中国的主流传统建筑，上到皇宫下到民宅都是院落式结构，中间是庭院，四周是房屋，与井田模式的相似性更强。相比北京的四合院，山西沁河的"4大4小"式堡院更能清晰地表现出"井"字结构，只是8份不再平等；中国南方家族聚居的大宅院，房屋就是围绕着一连串的"天井"布局，从平面结构上看就是一个或多个井字的组合，天井不仅仅是公共采光通风的渠道，也是公共

广东省大埔县泰安楼中的水井，旁边有井神的神位
The well in Tai'anlou, Dapu county, Guangdong Province. The well tablet is next to the well.

法国南特城堡中的水井，相比中国，欧洲人更重视水井的装饰
The well in Nantes castle in France. Comparatively, Europeans pay more attention to the decoration of the well than Chinese.

3. Focusing on the well

If Sijiaolou is one of the prototypes of the castle, then the Siheyuan is one of the most important Chinese residential prototypes. This prototype contains the most clear logic, that is Well-field System.

Well drilling technology in China, reportedly invented by Shen Nong. Humans learned to take water from drilling well is a significant event, afterwards, human farming and living do not have to over-rely on the river. It is not only much safer, but also brings much wider habitable range; most farming settlements are distributed around the well.

Chinese word "Well" in the oldest Oracle and inscriptions had the same basic strokes structure as in present. Two horizontal and two vertical the wooden frame at wellhead, in the seal characters, there is also a dot in the center. This graphic schema was later used to divide the land, gradually formed a living model, the social system, social order, cultural symbols, etc. A square land with a well in the center is divided by the "well" shape into nine portions, center portion with the well is public land, surrounding eight portions is equally distributed to eight families. Eight families in addition to farm their own lands, also co-cultivate the public land, the output of public land is for public reserves. This fairness, the spirit of collaboration, and the structured order (perfect order) obtained the favour of the rulers of the Zhou Dynasty. The Well-field is regarded as a land unit, also a social unit, and the basic unit for governmental taxations.

The Well-field system is gradually disintegrated in the late Chunqiu era, the most important aspect of Shang Yang Reform is the demolishing of the Well-field system, opening terraced rice paddies instead to accommodate large-scale agricultural production. However, the Well-field system has been described as an orderly social picture by a group of cultural patriots and idealists, Confucius as the representative, during the chaos of the Warring states. They determined to end the troubled times and trying to restore the social harmony. But after all, the Well-field mode is more suitable for clan- and tribal-scale population. As the scale of social organization expanded bigger, the Well-field relations of production hindered the development of productive forces even more. However, in serious social decadent confusion period, people have "the past is better than today" feelings, then the past would be excessive missed.

Well-field system carved a deep imprint in Chinese culture. Chinese people live in cities and towns regarded as living in the "marketplace" and leave their homes is termed as "displaced". The mode of the ancient towns and villages is more as the image of Well-field, a well as a public getting water place is usually located in the center of the towns and villages, around the wells there will be a square, the population living around the square; sometimes well is also placed in the village plaza. In short, because of the well will always

活动、劳作的场所。在这种建筑中，对弱化井田制平均主义，强化礼制所采取的方法是强化中轴线和正房尺度。

在初次看到闽南和潮州那种不合拢的弧段楼时，我们认为那是在山地等高线或水面岸线的制约下产生的，但再次仔细观察后看到，多数弧段楼并没有那些制约，与弧形有关系的是井，井是各个弧段的圆心，而且，许多井与弧段楼之间的地面上的鹅卵石铺地有放射状图案，很明显地表现出放射线是半径线。

更好的证实来自于椭圆形土楼形式中的几何逻辑性，在闽南平和县霞寨镇"雄山楼"这座椭圆楼中，尽管有一口井已被填了，我们仍能看到原来有两口井

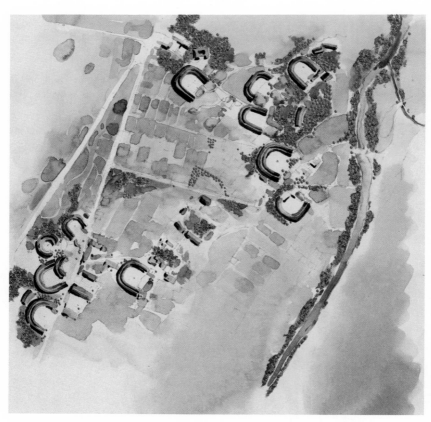

广东饶平县的弧段楼群，貌似围龙屋的外围，水井多处在圆心位置
The arc-segment buildings group, Raoping county, Guangdong. The external ring looks like a Enclosure Weilong House, wells are more likely placed in the centre of the circles.

form a public space, the public space will be the place of formation of villagers public will, thus becoming the representative symbol of the towns and villages.

The mainstreams of traditional architecture, from the imperial palace to the ordinary houses, are all courtyard-style structure. Courtyard is in the middle, surrounded by houses, very similar to the Well-field mode. Compared to the Siheyuan in Beijing, the "four large and four small" styled fort homes of Qinhe in Shanxi are more clear demonstration of "well" structure, only eight portions were no longer being equal. In the big houses inhabited by the Southern China family, housing is following a series of "courtyard" layouts. From the general plan, it is the combination of one or more "Well" characters. The patio is not just public lighting and ventilation channels, but also the place for public activities and crafts. In this type of architecture, the approach taken in the weakening of the field system egalitarianism and strengthening the ritual system is to emphasis the central axis and the size of the main house.

When the first time we saw the unenclosed arc buildings in southern Fujian and Chaozhou, we believed that they were under the constraints of the contour lines in the mountains or water shorelines, but after careful observation we realised that the majority of the arc buildings did not have those constraints. Well is linked to the arc, it is the center of all the arc segments. Moreover, the ground between the many wells and the arc buildings paved by cobblestone with radial pattern, it is clear indicated the radiation line is the radius.

It is better confirmed from the oval earth form of geometric logic as well. Inside the "Xiongshan Building", the oval building in Xiazhai town in Pinghe county in the south of Fujian, although there is a well has been filled, we can still see the original two wells are located in two of the center of the oval, therefore we can be sure at least how this oval building Xiongshan Building is created. First, dug two wells, then arranged the layout of all residential elements according to the centre of the two wells, so this oval building was produced.

After viewing a lot of the arc buildings and round earth buildings we had sensed that: round shape formed here are probably related to elevation contour lines, but by no means is the primary point, the primary point should be the well. Well here as center of a circle, round shape is fulfilled the requirement of each dwelling unit has equal distance from the wells. We imagine, when one or a few families started to settle here, the founding ancestor first to dig a well in an open space, and then roughly draw a circle or an arc using the well as the centre. The size of the circle or arc segment and the number of divided units is determined by the population and the number of households.

Emphasis on the defensive function, the arc closed to form the circle, and the external walls were heightened and thickened to form this particular castle, the round earth buildings.

分别位于椭圆形的两个圆心上，这样我们至少可以肯定雄山楼这座椭圆楼是如何产生的了，即挖了两口井，分别以两口井为圆心布局一个整体居住物，便产生了椭圆形。

看过诸多的弧段楼和圆土楼后，我们有一种感觉：这里的圆形之所以产生可能与等高线有关系，但那绝不是首要关系，首要关系是井，井在这里成为了圆心，圆形是每个居住单元要求与井等距产生的。我们想象，当年一个或几个家庭在此定居，开基祖先在一块空地上打到一口井，然后大致以井为圆心画一个圆或一个弧段，圆或弧段的大小及把圆弧平分成多少个单元由人口和家庭数决定。

强调防御性时，弧段合拢形成圆形，外墙加高加厚，就形成了圆土楼这种特殊的古堡。

难道在中国的东南边陲，那里的民居中也有井田制的原型，只是这种的井田制的图式不是方格形的，而是弧形、圆形的。如果这样的逻辑性能够确立，圆土楼的产生原因就好理解多了。

It is true that in China's southeast border region, where also have the prototype of the residential well-field system, only this Well-field system's schema is not a grid square, but an arc or circular shape. If this logic is able to establish, the causes of the round earth buildings creation more likely to understand.

饶平弧段楼前的水井
The well in front of the arc-segment building in Raoping county.

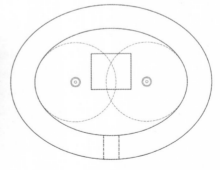

福建平和县雄山楼平面示意图，由两口水井形成的椭圆形
The plan illustration of Xiongshan Building in Pinghe county, Fujian. Two wells form an oval shape.

4. 漳州潮州交界处的历史文化特征

第一，那里的传统文化都有强烈的二元性，一方面极其强调礼制，民居的代表是所谓的堂横屋；另一方面是有同样强烈的原始性、自然主义，民居的代表就是弧段楼、围龙屋、圆寨、圆土楼。

第二，当地居民现在有闽南人、潮州人、客家人、畲族人之分，其来源主要是古代闽越人、南迁的中原汉人、畲民这3种，历朝历代中，都有畲民或畲族人融入闽南人、潮州人、客家人中。

第三，古代所谓的畲民是指主要居住在闽粤赣山区的、游动耕作、状态较原始的民族或人群。历史学家多认为，畲民源于荆蛮，初居湖南武陵源一带，与苗族、瑶族同源共祖。大约在唐代以前，畲民南迁至闽粤赣山区，唐代冯盎征的"獠蛮"、陈政父子开漳州时遇到的"溪蛮"、置汀州时开的"洞蛮"中应都有畲民成分，甚至是主体。这期间畲民一部分汉化，一部分进一步迁入深山或海边。宋代的文献中，对畲民多有记载：北宋《太平寰宇记》载宋代广东梅州地区的人口情况是"主为畲瑶，客为汉族"。宋末，元军中有一支福建畲军，元政府曾"令福建黄华畲军有恒产者为民，无恒产与妻子者编为守城军"。后黄华叛元，使"福建归附之民户口百万，黄华一变，十去四五"。以上情况说明在宋末元初时闽粤地区畲民众多。

畲民与羌人的情况很类似，他们都是迁徙人群，在古代人数众多，现在则人数极少，因为有大量人口不断融入汉族，现在的人口中有相当部分居住在古堡式的民居中。

第四，圆土楼不可能是南迁的中原人带过去的形式，因为中原人早就不习惯圆形建筑了。圆土楼应该是闽粤地区文化上更原始、更自然的人群首先采用的，这样的人群首要追求的是平等性，那么这样的人群应该是闽越人或畲民。

第五，潮州饶平县的凤凰山是畲民的一处"祖地"，凤凰山至福建平和县灵通山之间历来是畲民活动最集中的地区，同时那里也是弧段楼、圆土楼最集中的地区之一，而且，那里的圆土楼不乏历史悠久者，形式都比较原始，最重要的是，那里的古代民居几乎全部是弧段楼与圆土楼形式。

第六，上述地区的闽南人、潮州人、客家人、畲族人都住圆土楼、弧段楼；由于历史中有歧视问题，畲族人口曾经很少，但近年环境改善，许多曾经的客家人、潮州人、闽南人申请改回畲族身份。

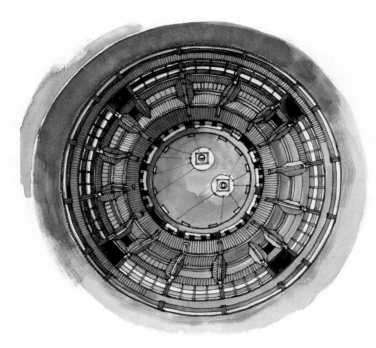

福建华安县二宜楼俯瞰图。虽然楼内有两口水井，但楼为正圆形，两口水井都有意识地闪开了十字形轴线，这种布局显示的成熟性在凤凰山土楼中非常罕见
The top view of Eryi Tower in Hua'an county, Fujian. Two wells place in this normal circular shaped building, deliberately avoid positioning on the two cross axes. It is a very rare layout in mature earth buildings in Mountain Phoenix region.

4. Historical and cultural characteristics in the border region of Zhangzhou and Chaozhou

First, the traditional culture has a strong duality in the region. On the one hand, great emphasis on ritual, residential representative is the so-called Tangheng House; on the other hand, there is equally strong original, naturalistic, the representative of the residential buildings is the arc-segment buildings, Weilong House, round village and round earth buildings.

Second, local residents are people of southern Fujian, Chaozhou, Hakka, She people, mainly from three groups, i.e. the ancient Fujian and Zhejiang group, the Central Plains Han migrated to the south, and the She people. Throughout past dynasties, the She people were kept joining in southern Fujian, Chaozhou and Hakka groups.

凤凰山土楼多是比较原始的形态
More native forms of earth buildings exist in Mountain Phoenix group.

Third, the ancient She people are the original groups living in the mountainous areas of Fujian, Guangdong and Jiangxi, shifting cultivation and in primitive living state. Historians agree that the She people from primitive states Jingman, the early living along the Wulingyuan region, has common ancestor and homology with the Miao and Yao minority groups. Just before the Tang Dynasty, the She people moved south to mountains in Fujian, Guangdong and Jiangxi provinces. The General Feng Ang fought and defeated Liao barbarians in the Tang Dynasty, Chen Zheng and his son encountered "Xi barbarians" in Zhangzhou, and confronted "Dong barbarians" in Dingzhou should have the She people involved partly or even the majority. During this period part of the She people converted into Han Chinese, part of them move further into the mountains or to the coastal areas. The literature of the Song Dynasty, has many records about the She people: Northern Song's "peaceful world" contained the record of the population in Meizhou area in Guangdong province in the Song Dynasty as "She Yao people are the host, Chinese Han are the guest". In the late Song Dynasty, there was a She Army from Fujian in the Yuan Army. Yuan government issued a policy stated that, amongst the personnel of Huanghua's She army, persons with inherited property would be the residents, persons with no inherited property or family would joined the city's defending army. Afterwards Huanghua rebelled against Yuan Government, the historic records noted that "accounts million of allegiance of China in Fujian, after the rebel, the population reduced down to 50%-60%." The above record indicated that there were many She people in Fujian and Guangdong regions in the late Song Dynasty.

She people and Qiang people are very similar, they both are migratory population, many in ancient times, rare in present. Because there are large numbers of She people continue to integrate into the Han population, a considerable proportion of the She group are living in castle-style houses now.

Fourth, the round earth buildings can not be the building style brought by the Central Plains people migrated to the south, because long time ago the Central Plains have already not used to the rotunda anymore. The round earth buildings should be firstly created by the primitive and native people in Fujian and Guangdong regions. It was important for the native crowd to pursuit of equality, such people should be local people of Fujian and Zhejiang or the She people.

Fifth, the Mountain Phoenix in Raoping County of Chaozhou is the She people's ancestral lands. It has always been the most concentrated areas of activities of the She people from the Mountain Phoenix to the Mountain Lingtong in Pinghe County in Fujian, where is one of the most concentrated areas of arc-segment buildings, round earth buildings. Also, the round earth buildings there is no lack of a long history, in more primitive forms, and most importantly, where the ancient houses are almost all in the forms of arc-segment buildings, as well as the round earth buildings.

Sixth, people of south Fujian, Chaozhou, Hakka, and She people all live in round earth buildings and the arc-segment buildings in the above areas. Due to the historical discrimination issue, She's population was very little, but in recent years with improved environment, many of the Hakka, Chaozhou, south Fujian people are seeking to return back to their She identity.

5. 凤凰山土楼的特征

第一，此地区所有古代民居都是圆形和弧形的，只有圆土楼，没有方土楼或方形碉楼式古堡，说明圆形、弧形是那里唯一的图形母题。如果这里的圆弧因素是这里的古人从外面带来的、学来的，为什么他们这么绝对？不引入多一点儿直线因素。这是否在说明，圆弧因素是这里原生的，是最根本的传统图式。

第二，各种圆形建筑种类齐备，有圆寨堡、圆围屋、独立存在的围龙和弧段、单环圆土楼、多环圆土楼、各种圆楼弧楼组成的原始状的村寨，甚至一座小城镇都基本是由各种圆楼弧楼组成。

第三，这里的圆土楼实际上多是多边形、不规则圆形的，那些土楼对比成熟型圆土楼时，会表现出许多圆土楼从初创到成熟的逻辑性。首先，所有圆土楼的内圈都是多边形的，因为将木材加工成弧形非常费力，将整体木构架作成正多边形也特别费力，只有木构架是正多边形，圆土楼才能作成正圆形。如果圆形只是为了表示公平性，就不需要太苛求圆形是否非常正规，那么圆土楼的外墙可以用十六边形、三十二边形等来代替圆形，也不用太在意圆形是否都是正圆形。所以，早期的圆土楼应该有常用多边形和不规则形的特点。而一旦圆土楼成为一种传统的形式、风格的追求，它就会以正圆形为主了，福建成熟型圆土楼集中区的情况就是这样。

第四，虽然此区域内的客家土楼与闽南、潮州、畲族土楼形式一致，但标准的客家土楼和非客家土楼在建筑空间构造上存在着一个显著差异，即二者的外形虽完全一样，但客家土楼的平面布局为单间房作为最小居住单元，由单边走廊并联，即通廊式；非客家土楼则为竖向一至数层先组成以小家庭为单位的最小居住单元后再并列一起，即单元式。单元式表现的是每户人家之间的平等性，这更符合井田制的原则。通廊式表现的是家族内部每个成员的平等性，这显然是后期的发展成果。

第五，井多位于圆心处或圆心附近，相比之下，客家的成熟型圆土楼由于格外强调正宗、秩序等，故需要强调中轴线，不仅大门对应的圆环上另一端的房间要设成祖堂，圆天井中间也多放祠堂，水井便只好靠边，所以我们看福建永定县、南靖县的客家土楼，很难建立起水井与圆楼的图形关系。

第六，相比之下不够强调祠堂、祖堂。

第七，多数圆楼、弧楼、圆围是不设防的，设防的只有少数高大的圆土楼和圆寨，表明这里的圆弧因素来源于民居，而非寨堡。

饶平的一组土楼建筑群,显示出土楼在一种强大的文化传统下既有控制,又自然的生长过程。这应该是一种原生形式的现象
A group of earth buildings demonstrated that architecture form changed under traditional culture influence and controls as well as natural evolution process.

5. Characteristics of earth buildings in the Mountain Phoenix

First, all the ancient houses in this region are circular and curved. There were the round earth buildings only, no square earth buildings or square Diaolou castles exist, that means circular or arc is the only graphic motif. If the arc factor was learnt and brought in from the outside by the ancients, why they are so absolutely? Does not introduce a little more straight-line factors? Does this mean that the arc factors is native and the most fundamental schema?

Second, various types of rotunda exist, e.g. round fortress, circle enclosure houses, stand-alone Weilong Houses, and arc segments, single-ring round earth buildings, multi-ring round earth buildings, round and arc buildings formed of primitive village, even a small town generally composed of a variety of round and arc buildings.

Third, the round earth buildings are actually more polygons and irregular shaped, Those earth buildings in comparison of mature round earth buildings will illustrate the logic of the round earth buildings evolved from start-up to maturity. First of all, all the inner circles of the round earth buildings are polygons, because the wood processing of arc shape is very laborious, and making the overall wooden frame into a regular polygon is particularly laborious. Only if the wooden frame is a regular polygon, round earth buildings can be made to a circle. If the circle is for showing fairness, it would not be necessary to restrict on the exact round shape, then the external walls of the round earth buildings can use 16- or 32- hexagonal instead of round, also do not be too concerned about whether the round form all are exact circular. Therefore, the early round earth buildings should have the characteristics of commonly used polygonal or irregular round. Once the round earth buildings developed into a pursuit of traditional form and style, it would be mainly based on the exact round form, the mature round earth buildings concentrated area in Fujian Province would be the case.

Fourth, although the Hakka earth Buildings in this area have the same form as those in the south of Fujian, Chaozhou, She group region, in comparison on the construction of building spaces, the standard Hakka earth building and non-Hakka earth building have a significant difference. Between the two shapes, although exactly the same, but the layout of Hakka earth building has a single room as the smallest living unit, connected parallely by the unilateral corridors, namely the Corridor style; non-Hakka earth building has the vertical space linked one to several storeys to firstly form the minimum living unit of a small family, and then tied together, that is called Modular style. Modular-style represents of the equality in each household, which is more in line with the principle of Well-field system. Corridor-style represents of the equality of each member in the big family group, which is obviously the results of development at a later stage.

第八，多环圆土楼都是以最内圈的高大圆土楼为中心生长出来的，表明最早建立高大圆土楼时主要是需要它的防御性，以之作为城堡使用，否则没必要把家建得那么封闭。凤凰山的圆楼虽多，但完全封闭的、古堡式的并不多，这符合常情，古堡在全部民居中的比例不可能太高，没必要的时候，人不愿总挤在古堡里生活。富裕的大家族始终住在古堡里是因为他们的古堡足够大，里面不挤。这里的家族在又需要盖新房时，如果不想再增加城堡数量，同时又想依靠着城堡，便可以在圆土楼外还以圆楼的圆心为圆点放同心圆，加建出可以是整圈、也可以只是弧段的新房子。这样就像前面介绍过的那些有大碉楼的设防村寨、围屋一样，可以平战结合。

第九，再一次强调，这里的古代民居几乎百分之百是圆形的、弧形的，而在成熟型土楼集中区的永定、南靖，圆土楼在各类古代民居中所占比重只有10%。

靠近漳州市的常规村落中孤立存在的小型单元式圆土楼
Isolated small modular round earth buildings in the conventional villages close to Zhangzhou City.

单元式土楼内部　Modular houses inside the earth building.

单元式土楼或弧段楼标准型单元的平面、剖面图。与内通廊式土楼一样，因为外围被分割进各个独立的房间，所以这种构造不是特别利于群体防御，这也说明土楼虽有防御性，但它最突出的并不是防御性，那么，它和军事堡寨的关系就不会是直接性的

The plan and section plan of a standard unit in modular earth or arc buildings. Similar to the corridor-style earth building, the periphery is divided into separate rooms, so this structure is not particularly conducive to group defense. It also shows that although the earth building has defensive nature, but the most prominent feature is not the defense, therefore, the design of earth building is not directly linked to military Fortresses.

Fifth, the wells were often located in the centre of a circle or near the centre of the circle. Comparatively, the Hakka mature round earth buildings pay extra emphasis on authentic and social order, so the axis is the focus. Not only the room opposite to the gate on the other side of the ring needs to be the ancestor temple, the middle of the round patio are mostly put the ancestral halls, wells had to move aside. So if we view the Hakka earth buildings in Yongding County and Nanjing County in Fujian, it is difficult to establish the graphical relationship of the wells with a round building.

Sixth, comparatively, no enough emphasis on ancestral temple, ancestral hall.

Seventh, the majority of a round building, the arc floor, round circumference are undefended, only a handful of tall round earth buildings and round village are fortified, that explains the arc factors are originated from residential houses, rather than the Fortress.

Eighth, multi-ring round earth buildings are built around the tallest round earth tower in the most inner circle, indicate that the first tall circle earth tower was mainly used as a defensive castle, otherwise there was no need to build a closed house. There are many round buildings in Mountain Phoenix, but many were not built completely closed as the castles. It is in line with common sense, the proportion of the castle in all residential areas can not be too high, unless necessary people do not want to always squeeze in castles to live. Wealthy family always lived in the castle because of their castle is large enough, with no crowding problem. When family needs to build a new house while still rely on the existing castle instead of building new ones, the houses can be located on concentric circles using the center of the round earth building as the reference centre point. The new houses can form an entire external ring or only contain few arc buildings, just like previous mentioned Diaolou of the fortified villages or round houses, they can be used in both peacetime and wartime.

Nighth, it is worth stress again once more that the ancient houses in this region (Mountain Phoenix) are almost 100% circular or arc shaped. However, in the mature form earth buildings concentrated areas, i.e. Yongding, Nanjing, the round earth buildings have only 10% amongst all kinds/types of ancient houses.

福建平和县的土楼，有多边形的痕迹
The earth buildings in Pinghe County in Fujian, with feature traced back to the polygon structure.

6. 大胆假设与小心求证

显然，我们在大胆假设凤凰山一带是圆土楼的祖地，其发明者很可能是在古代遭歧视，在近代仍然遭文化轻视的畲民。虽然我们已经尽力在求证，但由于能力问题，许多关键性要素之间的逻辑性还不能建立。比如，没有找到畲民更早期的建筑，也没有相关的文字记录，特别是有关"圆形井田制"的文字记录，凤凰山圆土楼的建造年代需要准确的考古鉴定等。

二、粤东土楼

粤东土楼集中在饶平、大埔两县。虽然数量庞大，但大多数已经破烂不堪，被混凝土的新民居搅得支离破碎，在地面上很难分辨其形状，特别是那些不规则圆形的土楼，只能靠卫星图像帮忙。

在古堡主题下，我们重点关注有古堡性质的土楼。

1. 饶平县道韵楼

规则八边形的道韵楼位于饶平县从前的县城三饶镇边，那里离凤凰山很近，虽然凤凰山的圆土楼许多实际上是多边形楼，但八边形还是不能混同于圆形，不过它进一步反映出凤凰山土楼造型的随意性，它虽然不是圆形、弧形，但与"圆

道韵楼正面　　The front view of Daoyunlou.

6. Bold assumptions and careful verification

Obviously, we boldly assume that the Mountain Phoenix area is the ancestral land of the round earth buildings, its inventor probably was the She people who was discriminated in ancient times and was still in the modern culture of contempt. Although we made every effort to verify, but due to limitations on time and resources, many of the key elements of logic could not be established. For example, did not find the earlier construction by the She people, there is no relevant written records, especially the written records of the circular Well-field system, the construction era of the round earth buildings in Mountain Phoenix requires accurate archaeological examination, etc.

Section 2 Earth Building in Eastern Guangdong

The earth buildings in eastern Guangdong are concentrated in Raoping and Taipo counties. Although there are a huge number still exist, most of them are already in decay. Fragmented by of the concrete house in new residential areas, it is difficult to distinguish their shapes on the ground, especially those irregular shaped earth buildings. Relying on the help of satellite images, the drawings and graphs are plotted based on google earth images.

Under the castle theme, we focus on the earth building with castle function.

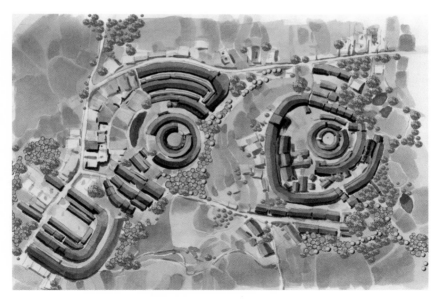

根据卫星图像绘制的凤凰山土楼俯视图
The perspective view of the earth buildings in Mountain Phoenix region based on satellite images.

道韵楼内部，多口水井分布在各单元附近
Inside of the Daoyunlou, several wells distributed around modular houses.

形井田制"的意识取向不矛盾。除了道韵楼，饶平还有几座八边形土楼，除此，其他地方极少见八边形土楼。

道韵楼据称始建于明代成化年间的1477年，其外墙高11.5米，比许多古城的城墙还高。其内切圆直径101米，如此大的楼在平日即可住几百人，有危险时足可容几千人。

道韵楼内的黄氏家族操潮州话，但自称是客家人。

2. 上饶镇镇福楼

上饶镇在三饶镇以北，是凤凰山圆土楼最集中的区域，据称始建于明代永乐11年（1413年）的镇福楼是凤凰山土楼的一个典型，它是一组由圆土楼、弧段楼组成的聚落的中心，其平面呈不太规则的椭圆形，楼内只有一口水井，位于中心位置，楼外还有一圈同心圆形式的弧段楼，没有合拢，以让出楼门前的广场和水塘的位置。其背后一座圆土楼基本是正圆形，水井基本位于圆心处，这座土楼建造年代比较晚，所以形状正规了。

1. Daoyunlou in Raoping county

Regular octagonal shaped Daoyunlou is located at the side of former Sanrao town in Raoping county, it is very close to the Mountain Phoenix. Although many of the round earth buildings in the Mountain Phoenix are actually polygon buildings, octagonal cannot be confused with the round shape, but it further reflects the arbitrariness of the shapes of earth buildings in Mountain Phoenix area. Even though the shape is not a circle or an arc, it is not contradict to the conscious of the circular Well-field system. In addition to the Daoyunlou, there are several octagonal earth buildings in Raoping, which are rare elsewhere.

Daoyunlou reportedly was built in 1477 in the Chenghua period in the Ming Dynasty. Its external wall is 11.5 meters high, even higher than many of the ancient city walls. The diameter of its inscribed circle is 101 meters, so hundreds of people can live in the building during peacetime, it can accommodate thousands of people during chaotic time/wartime.

The Huang family living in the Daoyunlou speak Chaozhou dialect, but they claimed to be Hakka.

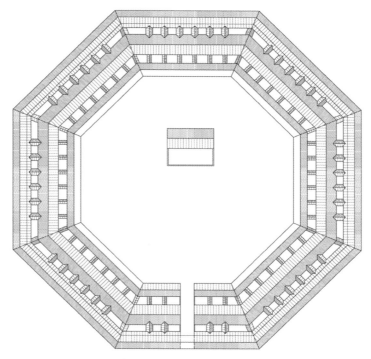

道韵楼平面示意图　　The illustration of the plan of Daoyunlou.

2. Zhenfulou in Shangrao town

Shangrao town in the north of Sanrao town, is the most concentrated area of the round earth buildings in Mountain Phoenix. The Zhenfulou is a typical earthen building in Mountain Phoenix allegedly was built in Yongle 11th year (1413) in the Ming Dynasty. It is the center of the settlement formed by a group of the round earth buildings and the arc-segment buildings. The plane of the building was irregular oval, inside the building there is one well located in the centre. Outside the building there are arc-segment buildings on the concentric circles, these arcs are not closed to provide spaces for the plaza in front of the door and pond. Behind this building there is a round earth buildings in basic round shape, a well is located almost at the center of the circle. As this earth building was constructed relatively late, so it has the regular shape.

根据卫星图像绘制的镇福楼建筑组群俯视图
The top view of Zhenfulou building group, drawing is based on satellite images.

3. 上善镇南华楼

上善镇在上饶镇以北,镇外南华楼的独特处在于,它的平面格局为前方后圆,井对于后部半圆圆心的位置只是偏离了一点儿,以闪开大门至祖堂的中轴线。

3. Nanhualou in Shangshan town

The Shangshan town is in the north of Shangrao town, outside the township the Nanhualou has unique feature with rectangular shape in the front and circular shape in the rear. The well is located slightly off the centre of rear semi-circular to deviate from the axis through the door and the ancestral hall.

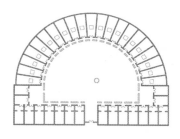

南华楼平面图　　The plan of Nanhualou.

上4图：福建南靖县田螺坑土楼群,虽然其中的几座圆土楼建造年代较晚,但由于这几座土楼都是实际居住性土楼,所以它们的水井大致在圆心处

4 photos above: the earth building group in Tianluokang region, Nanjing county, Fujian Province, there are still people living inside, the well is located in the center.

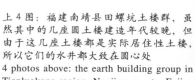

三、福建圆土楼

不论圆土楼的源头究竟在哪里，综合质量最高的一批圆土楼显然多在福建。

1. 漳浦县锦江楼

最有古堡意味的圆土楼应该是漳浦县深土镇锦东村的锦江楼，它所处的位置相对靠海，远离圆土楼的几个集中区，但是楼主人陈氏家族是在清代"复界"以后从客家地区迁来的，应该带来了成熟型圆土楼的设计，同时，这座土楼也有一些创新设计。

土楼成熟区的圆土楼
A round earth building in the mature-styled region.

成熟区里也有不规则的土楼
Irregular shaped earth buildings also exist in the mature-styled earth region.

Section 3 Round Earth Buildings in Fujian Province

Regardless of the originations of the round earth buildings, the bunch of round earth buildings with the highest overall quality are clearly found more existence in Fujian.

1. Jinjianglou in Zhangpu County

The most significant round earth building with castle features should be the Jinjianglou in Jindong village in Shentu town in Zhangpu County. It's located relatively near the sea, away from several round earth buildings concentrated areas. The Chen family moved from the Hakka area after the governmental "restoring coastal regions' campaign" in the Qing Dynasty. They should have brought the mature-form deign of round earth buildings, at the same time, this earth building also implemented some innovative design.

The Jinjianglou has three rings of buildings. The inner ring, a three-storey circular tower, was built in the 56th of Qianlong year (1791) in Qing Dynasty; the middle ring was built in the 8th year in Jiaqing era (1803) in the Qing Dynasty, is also circular, 2-storey; the outer ring is circular but not closed, with gaps for showing the doors of inner and middle rings. Although the three rings of buildings were built in sequence with no unified planning in advance, still the final completed buildings perfectly stood together and complemented each other seamlessly.

锦江楼正面   The front view of Jinjianglou.

锦江楼的内环和中环之间
View between the inner and outer rings of Jinjianglou.

锦江楼共有内、中、外三圈建筑，内圈楼建于清代乾隆五十六年（1791年），正圆形，高3层；中圈楼建于清嘉庆八年（1803年），也是正圆形，高2层；外圈也是正圆形，但不合拢，闪开了中圈、内圈的大门，高1层。虽然3圈楼是先后建的，而且事先没有统一规划，但最终楼完美地浑然一体。

楼的3圈布局类似凤凰山的多环式土楼，但凤凰山的土楼只突出主楼，外环楼无论几圈，都一样高，不会步步高。楼的内、中圈有高出楼面的门楼，则是借鉴了漳浦当地古堡的形式，门楼高出楼面不是只为造型，而是使门楼中的楼梯通至楼面的走马道，这样，土楼的外墙更接近于一般古堡的城墙。为了这种构造，锦江楼除了外圈的屋顶是双坡顶，中、内圈都是单坡顶，这一点与其他圆土楼都不同。总之，锦江楼设计灵活，既是土楼，也是一座有双层城墙的古堡。

2. 平和县椭圆形砖楼

平和县霞寨镇钟腾村有一座很少见的外包灰砖的椭圆形土楼，楼建于清代乾隆年间，楼主人黄氏家族曾在7兄弟中出了4名进士，有一人还高中榜眼，为此，家族特意在土楼所在的山坡下建了一座院落式的榜眼府，看来，黄氏家族认为土楼还不足以表达家族的新成就，至少不是高级对外礼仪场所。那么当初他们为什么建土楼？要么是为了家族安全，要么是他们认为家就应该是土楼。

楼位于溪边的半山坡上，楼的大门没有开在椭圆形的两个轴线上，而是斜向地开在一角，入门后才能看到，楼有两圈，内圈围楼的入口与外层门洞的轴线是错开的，楼内中心处有一口已废的圆井，该井位置并非椭圆形两圆心的一个，而是椭圆形长短轴的交叉点。沿短轴方向，此井对应着一座小祖祠，祖祠的前部呈梯形，似在对应圆形的放射线，内圈楼在祖祠前断开，以留出祖祠的位置。

该楼最为精彩之处在于其设计者经过几次巧妙的变轴，将入口与对应楼后

The 3-ring layout of the earth building is similar to those polycyclic earth buildings in the Mountain Phoenix, but the earth buildings in Mountain Phoenix only highlight the main building, the outer ring buildings, regardless of laps, all have the same height, with no step changes. The gatehouse of inner and middle ring rose higher above the floor levels, with reference to the form of local castle in the Zhangpu area. The elevated design is not only for the shape, but also to make the staircase in the gatehouse to join the bridleway on the floor level, therefore, the external wall of the earth building is very similar to walls in the general castle. To fulfill this design intention, the Jinjianglou has double pitch roof on the outer ring building, and single pitch roofs on both the inner and middle rings, which is very different from other round earth buildings. In summary, the Jinjianglou expresses the design flexibility, it is an earth building, also a castle with double-layer walls.

2. Oval brick building in Pinghe county

In Zhongteng village in Xiazhai town in Pinghe county, there is a rare oval earth building with gray masonry façade. It was built in the Qianlong period in Qing Dynasty, the landlord Huang family had four out of seven brothers to be official scholars, and one of the four was even ranked as the second place/(runner up) in the official/governmental exam. Then the family specially built a courtyard-style House for the second place winner under the hill

灰砖土楼的外围　The periphery of earth building with grey bricks cladding.

镇山龙脉的椭圆短轴方向连接起来，楼门偏向一侧躲开了溪边的断崖，内外门洞错开避免了楼内被一览无余，辅以抹圆角的细节处理，使内门不因变轴而显得逼塞。该楼虽不很大，但处处显出匠心，它处在从凤凰山土楼中心到客地土楼中心的半路上，像个里程碑，标志着土楼的设计思维越来越缜密。

从大门望内圆土楼　View from the gate to the inner ring of the earth building.

大门洞里的神位　The ancestral tablet inside the gate.

that the earth building stands. It appears that the earthen building of the Huang family could not adequately express the family's achievements, in another word, it was not regarded as decent social and ceremonial sites. But why they built the earth building originally? Either it was built for family safety, or they thought that home should be in an earth building.

The oval house is located halfway up the hill next to a stream. The gate of the house was not placed on the two axes of the elliptical, but it opened obliquely in the corner. After entering the gate, we can see the house has two rings, the entrance of the inner ring are staggered from the axis of the outer ring door openings. An unused round well is in the centre of the building, the well's location is not on either the elliptic centers of the oval, but on the cross point of the ellipse long and short axes. Along the short axis direction, this well is facing to a small ancestral hall, the front of the ancestral hall has a trapezoid shape, like on the corresponding radius lines of circular. The inner ring building disconnected in front of the ancestral hall so that allows the space for placing the ancestral hall.

The most wonderful highlight about the building is that its designers cleverly applied several changes of axes. The entrance of the building was linked with the dragon venture of dominating hill along the ellipse minor axis direction, the doors of the building? to one side to avoid facing the cliff near to the stream, inside and outside openings staggered to avoid see through into the buildings, supplemented by design details on the filleted corners, therefore, the inner door is not appeared to be constrained by the variations of axes. Even though the scale of this building is not huge, but always shows the ingenuity in design and creativity. It stands on the half way from the centre of the earth building area in the Mountain Phoenix to the centre of Hakka earth buildings area, like a milestone, marking the earth building design becomes more and more sophisticated.

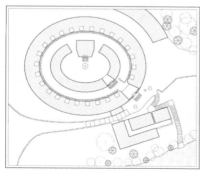 

土楼平面图　The plan of the earth building　　中心的水井　A well in the centre of the building.

3. 绳武楼和丰作厥宁楼

平和县芦溪镇以北,就是成熟型土楼集中的南靖县和永定县,所以芦溪的土楼形式就开始正规起来。镇中的绳武楼是最精美的圆土楼之一,首先楼前有环境设计,土楼前一般有半圆形的水塘,绳武楼门前有溪流就不再需要水塘,它用一排由茂密的竹子、杉树组成的林带将楼门与溪水隔开,楼门是歪的,这些都是风水学的影响。入楼后,满眼的精美木雕令人眩目,土楼里很少有这些精美的装饰。楼中圆天井里有一口水井,不在圆心上,但离圆心不远。

镇中另一座名为"丰作厥宁"的大楼因遭过火灾主体不太完整,但它的巨大感反而因此增强。楼内圆心处有一座由三眼井形成的大水井,井台的石头上布满沧桑。圆楼的外围还有一圈不合拢的弧段楼,在土楼大门至溪流边形成一块前庭状的空间,里面还有一组祠堂式的建筑和小神庙,溪边有一棵巨大的古树,建筑空间与自然环境结合得与绳武楼一样精彩。

绳武楼内部　Internal view of Shengwulou.

绳武楼斜向的大门
The diagonally aligned gate of Shengwulou.

丰作厥宁楼的复原鸟瞰图
The bird's eye view of reconstructed Fengzuo Juening Building.

丰作厥宁楼的入口巷道
The street path from the gate of Fengzuo Juening.

丰作厥宁楼内部
Internal view of Fengzuo Juening Building.

丰作厥宁楼的水井　　The well inside the Fengzuo Juening Building.

3. Shengwulou and Fengzuo Juening Building

In the north of Luxi town in Pinghe County, Nanjing County and Yongding County are the concentrated areas of the mature form earth buildings, then starting from Luxi, the form of earth building began formal up. Shengwulou in the town is one of the most beautiful round earth buildings. First of all, it has landscape design in front of the building, normally a semi-circular pond is placed before the earth building, Shengwulou building has stream running in front of it, the pond is no longer needed. It has a band of dense bamboo and cedar forest to separate the door from the stream, influenced by *feng shui*, the door of the building is crooked. Into the building, stunning wood carvings dazzled the eyes, few of the earth buildings were beautifully decorated like these. A well is placed in the round courtyard, not the center of the circle, but not far from it.

Another building in the town is called "Fengzuo Juening" building has no completed main body left after fire, but in contrary the sense of its large scale was therefore enhanced. A large well formed by three wells is in the center of the inner ring/circle, where the stones were covered with the vicissitudes of life. The periphery of the circular building is a ring of un-joined arc-segment buildings. From the door of the earth building to the stream side a vestibule-like space is formed, there is also a group of ancestral temple-style building and a small temple, there is a giant ancient tree next to the stream. The harmony of building space and the natural environment has created the same wonder like the Shengwulou.

4. 永定县承启楼

永定县高头乡是有"圆楼之王"称誉的承启楼的所在地，其实承启楼远不是最大的圆楼，它直径是73米，而饶平镇福楼的椭圆形短边也有80米，长边近100米。不过，承启楼构造复杂精致，誉之为王亦不为过。楼主江姓家族据称是早年为避难从江西迁来此地，大楼请当时江西有名的风水师设计，明末开工，至清康熙年间竣工，三代人花了半个世纪的时间才建成这座4环4层的大楼。

与凤凰山土楼和漳浦锦江楼不同，承启楼是外圈楼最高，次一圈为2层，最内部两圈楼都是1层。最内一圈是公共会堂，它最初的用途应是祖堂；第二圈是连排的一进小院，供会客、族内少年读书使用；第3圈亦为公共用房；外圈4层，首层是厨房等，上层才是居室。

承启楼内的居民们认为，他们的大楼是按太极八卦设计的，中心的公堂为太极，东西两口水井为两仪，外圈楼中4道封火墙将楼分成四相，每相中有楼梯间再分成八卦，每卦中8个房间便成64卦。

承启楼外部，与其他圆土楼形成群体时更有古堡意象
The exterior of Chengqilou. It shows more castle form when grouped with other round earth buildings.

俯瞰承启楼组团　The perspective view of Chengqilou building group.

承启楼内部　Inside of the Chengqilou.

4. Chengqilou in Yongding County

Chengqilou, so called the "King of the Round Building" is located in Gaotou town in Yongding County. In fact, Chengqilou is far not the biggest round earth building, which is 73 meters in diameter, while Zhenfulou in the Raoping town has the oval short axis of 80 meters, and the long axis of 100 meters. However, the Chengqilou has complicated and delicate structure, which was well deserved as the "King". The owner Jiang family is alleged to come from Jiangxi escaping conflicts in early years, the design of building was done by the famous feng shui master at the time. The construction started in the late Ming Dynasty and completed in Kangxi years, three generations have spent half a century before the completion of this giant 4-ring and 4-storey building group.

Different from the earth building in Mountain Phoenix and the Jinjianglou in Zhangpu, Chengqilou has the tallest outer ring, the second ring has 2-storey and two inner rings only has 1-storey. The most inner ring is a public hall, it initially used as the ancestral hall; the second ring has the row of small couryards for the parlor, family, juvenile studying; the third ring is also a public space; The outer fourth ring has the ground floor used as the kitchen and other utilities rooms, only the upper floor was? the bedrooms.

The residents of the Chengqilou believe that their building is designed according to the *Taiji* and Eight Diagrams principles. The central public court as the *Taiji*, the two wells in the east and the west are the two poles, the firewalls in the outer ring divide the building into four sections, each section have stairwells into eight segments, and then eight rooms in each segment form the 64 hexagrams.

方圆土楼之间的夹缝
The gap between rectangular and circular earth buildings.

5. 洪坑村振成楼

高头乡西面的湖坑镇洪坑村是一座几乎全部由土楼组成的村落，不过村中主要是方土楼，圆土楼只有几座，但其中有"圆楼王子"振成楼。

振成楼建于 1912 年，那时，建土楼已经基本上没有防御意义了，只是传统文化、习惯和风格爱好使然，客家人则仍然认为，土楼这种形式最能象征大家族的团结，但富裕的大家族不会再挤在一座土楼里，他们这时建土楼就如同建别墅，也如法国文艺复兴时期、欧洲浪漫主义时期那些按照古堡形式建庄园、别墅的行为一样。

洪坑村林氏家族靠烟草业致富，那时，科举制已废，这种家庭开始让子弟出洋留学，振成楼的设计建造者就是一位归国留学生。在客家成熟型圆土楼中，振成楼可说是设计最精巧严谨的一座，如和平县林寨的谦光楼一样，采用整体设计方法。从外部看振成楼并没什么特别，与众不同之处只在于中轴线垂直的两侧门外有两个弧段型院落建筑作为侧翼，似乎是在拱卫两个侧门，又有些凤凰山土楼边的弧段楼的意味。内部设计则将太极八卦、厅堂轴线序列、公私空间分隔、单元式和通廊式构造等诸多土楼设计要素有机地统筹起来。

圆土楼共三个外门，正门南门为天门，东西两侧门分别为地门、人门，据说过去天门平时不开，故地门、人门所在的横轴得到强化，楼内两口水井也设在此二门的门厅后天井中。正门天门似乎是在后天八卦图的乾位上，由此至楼后巽位的中轴线上分布着四厅三天井，位于中天井后的高大中堂的设置不仅有效地强化了圆形中中轴线的存在和厅堂空间序列，也使中心圆天井产生中庭的效果而不再只是一块场地，中堂也相应地产生了舞台的效果，事实上，中堂除可作为客厅外，本是可做戏台使用的，看台位置除中庭外还有内环楼的两层外环廊。中堂还特别设计成西式风格，正面像座希腊神庙，突出了舞台的效果，开敞的中堂和内环楼的敞廊亦使中庭空间的尺度和形态更为宜人。

以中庭为中心大楼共有两层环楼，二层内环楼除有看台作用外房间均为公共厅堂，外环楼高 4 层，平面上由封火墙分成 8 份，以对应八卦布局，其中三门和祖堂部分分别是乾、坤、巽、艮位，房间为公共用途，其余的兑、离、震、坎四位是四个半封闭式的居住单元，有各自的楼梯，因地坪层上两道环楼间的露天环廊在纵横轴线上有墙体和敞廊分隔以突出轴线空间，故此露天环廊被分割出四个相对独立的弧段型天井对应着 4 个居住单元，八道封火墙上又有门可开可合，使振成楼的居住空间兼具了单元式和通廊式两种形态。

5. Zhenchenglou in Hongkeng Village

Hongkeng village in Hongkeng town in the west of Gaotou town is a village almost all composed by the earth buildings. There are mainly square earth buildings with only a few round earth buildings, including the "Prince of the Round House"—Zhenchenglou.

Zhenchenglou was built in 1912, by then building earth house basically has no defense significance, it only dictated by the traditional culture, habits and style preferences. The Hakka people still believe that the earth building form is a symbol of the unity of the large family. A wealthy family will not squeeze in an earth building anymore, from then on they built earth buildings just like building villas, which is very similar to the trend of building manors and villas in the form of castles in the French Culture Renaissance and the European Romantic periods.

The Lin family in Hongkeng village relies on the tobacco industry to get rich. While the imperial examination system has been abolished, the rich families began sending their children abroad to study. The architect of Zhenchenglou house was a returned overseas student. Amongst the Hakka mature round earth buildings, the design of Zhenchenglou was the most sophisticated and rigorous one, it also applied the overall design approach, just

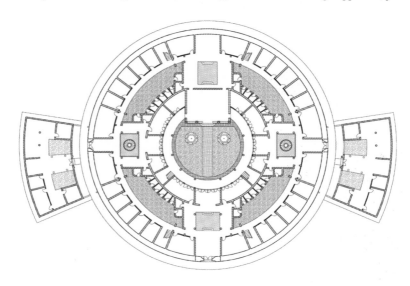

振成楼平面图
The plan of Zhenchenglou.

楼内轴线上的水井　A well placed on the axis of the building.

like the Qianguanglou in Linzhai village in Heping County. The exterior of Zhenchenglou did not show anything special, its uniqueness lies in there are two arc-shaped courtyard buildings extended as flank from both side doors perpendicular to the central axis. They seem to surround the two side doors from the outside, but also imply the feature of the earthen arc-segment buildings in the Mountain Phoenix. Many of the earthen interior design elements, e.g *Taiji* and Eight Diagrams, sequence of hall axis, separation of public or private space, modular and corridor structure, were organically co-ordinate up.

There are three external gates in this Zhenchenglou round earth buildings. The main entrance of the south gate is regarded as the Heavenly Gate, east and west sides entrance is the earth gate and the man gate respectively. It is said in the past the heaven gate is usually not open, consequently, the horizontal axis through the earth gate and the man gate being strengthened, also two wells located in the courtyard foyer behind each of the two gates. The main entrance gate seems to be acquired Qian Place in the Eight Diagrams, the central axis from here to the rear of the building distribution of the four halls and three patios, and the end of the axis is at the Xun Place of the Eight Diagram. The location of the giant Main hall which behind the middle patio set not only to strengthen the axis in the sequence of the existence and hall space to make the patio of the center circle in the atrium effect is no longer just a venue, but also to produce the effect of the stage. As a matter of fact, in addition to be used as living hall, the central hall can also be a stage. The inner building and the two storey outer ring's corridor can be used as grandstand, so does the central court. The central hall also specifically designed with the Western style, the facade is like a Greek temple, highlighting the effect of the stage, the airy central hall and the loggia in the inner ring exhibit the scale and shape of the atrium space pleasantly.

There are two-layer of ring buildings surrounding the atrium, in addition to the function as viewing platforms/halls the rooms are used as public halls. The outer ring 4-storey building is divided into eight sections by the firewalls corresponding to the Eight Diagrams layout, of which three doors and ancestral hall is occupied the *Qian, Kun, Xun, Gen Place*, which represent heaven, earth, wind and mountain, the rooms are for public use, the rest was four semi-enclosed living unit which has its own staircase, occupied the *Dui, Li, Zhen, Kan Place* which represent lake, fire, thunder and water. In the ground layer the open veranda between the two ring floor separated by walls and loggia to highlight the axis of the space, therefore the open-air react building is divided into four relatively independent of the arc patio corresponds to four dwelling units in the vertical and horizontal axis. The door on each of the eight layers' fire-wall can be either opened or closed, both the modular and the vestibule are two forms of the living space in Zhenchenglou.

楼内中庭一侧的舞台
The stage on central courtyard side.

楼内中庭一侧的看台
The viewing platforms on central courtyard side.

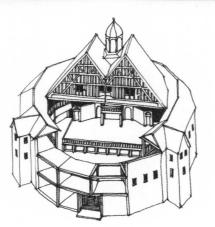

对比欧洲的古代剧院建筑
In comparison with European ancient theatre's architecture.

第五章 文化融合的大古堡
Chapter V Culturally Integrated Castles

广东新丰县儒林第的大望楼
Watchtower in Rulindi, Xinfeng County, Guangdong.

一、古堡文化的集成

1. 南北、土客、华夷文化的融合

事实上,我们前面讲过的洪坑村振成楼、林寨谦光楼、始兴满堂围等已经是一种文化、风格、形式综合性的民间古堡建筑,它们是中国南北方文化融合的产物,这种文化融合在千百年间通过贸易、战争、人口迁徙等不断在进行。

古代的中原人历来有文化优越感,他们迁到南方以后,必然坚持中原文化,但也不可能不受到南方土著文化的影响。而随着中国经济文化中心不断南移,同时中原文化不断被注入游牧文化成分,使得早期所谓正宗的中原文化反而在南方比在北方保护得好,加之南方的经济优势逐渐带动文化优势,使南方文化开始影响北方,山西沁河古堡的设计,据说就受到过南方风格的影响。

南方古堡最密集的地区是客家地区,在固守中原文化方面,客家人的态度最积极,客家古堡的名字都出自《经史子集》,但客家文化也吸收了大量闽粤赣地区的土著文化,风水学可谓是客家文化的一个重要内容,它虽然在中原文化中固有,但它是在闽粤赣地区成熟的,然后影响北方。

有南洋风格的开平碉楼　Kaiping Diaolou of the southeast Asia style.

Section 1　Integration of Castles into Culture

1. Cultural Integration between: North and South, the natives and the Hakkas, the Han people and the people of various minorities

In fact, the previously discussed Zhenchenglou in Hongkeng village, Qianguanglou in Linzhai, and Mantangwei in Shixing County are all comprehensively culturally and stylistically representative of vernacular castle architecture. They are products of the ongoing fusion of northern and southern cultures of China, though trade, war, and migration, over thousands of years.

The ancient people of the Central Plains have always felt superior culturally. Even after having migrated to the southern parts of China, they still adhered to the culture and traditions of their home, but were, nonetheless, influenced by southern culture. With China's economic and cultural centers moving south, the Central Plains culture was constantly being injected by aspects of nomadic culture, making the authentic culture of the Central Plains a shadow of what it once was in the Central Plains and much better preserved in the south than in the north. The south's economic advantages coupled with the increasingly superior culture in the south slowly began to influence the culture in the north. The Shanxi Qinhe Castle is said to be influenced by southern styles.

The most densely populated areas of the castles in the south are areas of the Hakka people. In sticking to the culture of the Central Plains, the Hakka people were very active and named their castles after Confucian classics and history. The Hakka people also absorbed a lot of the indigenous culture from their homes in Fujian, Guangdong, and Jiangxi. *Feng shui* can be described as an important element of the Hakka culture, but while it's present

福建永安市绍恢堡　　Shaohuibu Castle in Yong'an city, Fujian.

南方人下南洋的情况早至宋代就有，清代规模空前，清中期开始，在南洋致富的侨商纷纷回乡建房，他们将南洋建筑风格带回家乡。

2. 古堡设计的成熟

这种融合的文化至清代开始，在闽粤等地造就出一批综合形的大古堡，这些古堡在建造前显然都有完备的规划设计，其建筑师就是中国古代的风水师，他们首先从环境入手，使建筑布局符合风水学的要求，并有相应的构造配置；建筑中心部分结合礼制和居住功能，外围强调防御性，后部往往是风水要素最集中的地方。

二、闽中大堡

1. 三明市莘口镇松庆堡

福建的这种大古堡集中在闽中的三明市，当地人称之为土堡。莘口镇的松庆堡建于清嘉庆年间，平面总体呈圆形，前部抹直。它选址在一座坡度达20°以上的山坡上，只作为防御体的外围顺坡而建，整体逐步升高，后半部分因地势而升高剧烈。围内开辟为三层台地，上面布局院落式建筑。整体建筑外形奇特，介乎于圆土楼和围屋之间，外围走马廊上的坡屋顶层层叠叠升高的形态应该有风水意义。

2. 水美3堡

沙县南部的水美村中现存3座土堡，相传为从闽南迁来的张氏三兄弟在此经营茶叶发财后所建，3座堡分别名为双吉、双兴、双元。

3座堡中以双元堡保存最为完整，它也最古老，为清代道光年间所建。其外形前方后圆，比松庆堡规整，方圆之间衔接自如，但各自的轮廓清晰，外围墙高大，呈更强烈的古堡形态。

建筑整体沿中轴对称，前脸的两端各有一座凸出的角楼，其屋顶的起脊方向与前脸部分垂直，由此，两侧外围的屋顶脊向一致同时向后层层抬高，靠近后部时开始抹圆，圆弧正中为最高点，设小望楼一座，与角楼一样，小望楼也突出于外围墙体。外围墙体分两层，一层是厚达4米的实体墙，二层实体墙只剩外侧的1米厚度，内侧为木结构环廊，高度比内部房屋的二层空间要高，当

in the Central Plains, it matured in the south and later reappeared in the north to affect the culture there.

Chinese settlement in Southeast Asia began as early as the Song Dynasty, grew to an unprecedented scale during the Qing Dynasty. In the middle of the Qing, the Chinese businessmen in Southeast Asia began returning north to build houses after they made it big, bringing back with them the architectural style of Southeast Asia.

2. Maturation of Castle Design

This integrated culture began in the Qing Dynasty, and in Fujian, Guangdong, and etc., were built a number of large, culturally integrated castles. These castles were clearly designed and planned by an architect who had mastered the art of *feng shui*. Beginning with the environment, the layout of the building must meet the requirements of *feng shui* with corresponding structural configuration; Centralized is the combination of the ritual traditional and the residential, surrounded by heavy defenses, in the very back of the castle lie most of the *feng shui* elements.

Section 2   Great Castles in the Middle of Fujian

1. Songqingbu Castle in Xinkou Town, Sanming City

Such large castles in Fujian are concentrated in Sanming City, known locally as the earthen castles. Songqingbu Castle in Xinkou town was built during the reign of Emperor Jiaqing. It has a circular shape overall, but is flat in the front, and is located on a slope greater than 20°. The wall surrounding the castle was dictated by the terrain and rose with the mountain with the back being much higher due to the added height provided by the mountainside. The interior of the castle is split into three tiers, rising with the mountain, on each of which was built a courtyard with surrounding buildings. The peculiar shape of the castle lies somewhere between that of round earthen buildings and enclosed houses, and the rising layers of covered porches are significant to *feng shui*.

2. Three Castles in Shuimei village

In southern Shaxian County in the village of Shuimei exist three earth castles. According to legend, three Zhang brothers from southern Fujian built them after they made a fortune from selling tea. The three castles were named: Shuangji, Shuangxing, Shuangyuan.

The best preserved of the three is Shuangyuan. It is also the oldest, built during the reign of Emperor Daoguang during the Qing Dynasty. Its shape is half-square half-circle, square

松庆堡正面　The front view of Songqingbu Castle.

松庆堡侧面　The side view of Songqingbu Castle.

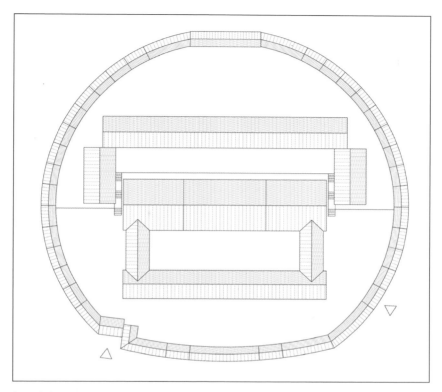

松庆堡平面图　The plan of Songqingbu Castle.

双元堡正面　The front view of Shuangyuanbu.

环廊与内部房屋的二层空间相通时，就需用楼梯连接，且楼梯呈凌空斜架状，在民居中出现此种景象实在少见。

纵观整个土堡，其构造精致完整，空间复杂而有序，实用而有趣，反映出很高的设计境界，据称，它是按清朝福州官办的设计局所提供图纸兴建的，这一点提醒我们不能轻视古代的建筑设计体系，不要认为民居都是匠人凭经验建成的。

双兴堡也是前方后圆的土堡形式，但构造与双元堡有所不同，其外围前后部分均有塌毁，以现状来看，其内部比双元堡小而简单，但其外围构造更复杂。外围后部的这种弧形因素应该是风水学影响的，但在也讲究风水学的北方民居中，几乎没有弧形因素。

双吉堡虽然称"堡"，但没有什么防御性。

双元堡的屋顶　The roof of Shuanyuanbu.

双元堡内上外围二层的楼梯，外围二层为贯通的环道，比圆土楼利于防守
The staircases reach the top level of the external ring wall. The top level ring has through pathway, easier for defense than round earth building.

in the front and round in the back, with a smooth transition between the two. The tall outer wall conveys an image of a strong castle.

The castle is symmetric about the middle axis running front to back. On both corners of the front protrudes a corner tower outwards. The roofs of the outer wall layer up towards the back of the castle, all the way to the center of the back wall. There, the highest point of the outer wall, a watchtower sits along with a corner tower that also protrudes from the wall out the back. The outer wall consists of two stories, the first of which is a 4-meter thick solid wall. The second story is a covered causeway that has a 1-meter thick wall aligned with the outside of the wall on the first floor. This causeway is higher than the second stories of the houses within the castle and the two are connected by suspended stairways, which is rare in residential areas.

Throughout the earth castle, the construction is intricate and wholesome, the space, complex but orderly, practical but fun. It reflects a high degree of design innovation. It is claimed that the design bureau of the city of Fuzhou provided the construction drawings during the Qing Dynasty. This reminds us not to underestimate ancient architectural design, or to assume that all vernacular buildings are spawned from pure experience of their craftsmen.

双元堡侧后面　The side and back view of Shuanyuanbu.

Shuangxingbu Castle is also square in the front and round in the back, but its structure differs from that of Shuangyuanbu Castle. While it is apparent that its interior is smaller and simpler than that of Shuangyuanbu Castle, the periphery, which has since collapsed, appears more complex, with arcs influenced by *feng shui*, but there are no any arc elements in northern residence architecture which also constructed according the rules of *feng shui* system.

Shuangjibu is called "Castle", but there is nothing defensive about it in nature.

双兴堡正面　　The front view of Shuangxingbu.

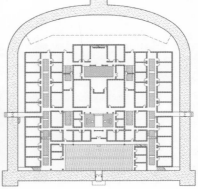

双元堡平面图　　The plan of Shuangyuanbu.

双吉堡正面　　The front view of Shuangjibu.

3. 永安市安贞堡

著名的大土堡安贞堡坐落于槐南镇外，其构造与水美的双元堡相近，据说也是根据完整的设计图纸施工建造的。它为池姓家族建于清代同治年间，选址在一个山坳的尽处，因借山势，加之自身高大的体量，又面对宽阔的山间盆地，故显得极其宏伟壮丽。在外形上，它与我们前述的土堡不同。第一，它前面左右两侧的角楼屋顶为四角攒尖式；第二，其后部中间望楼的屋顶上又向后挑出一间小阁楼，这两点与汉墓中出土的四角楼明器有相同之处。

整体来看，堡内中心构造为一独特的方形围楼，构造复杂，但有条不紊，然安贞堡最精致的部分还要数它的外围，其外围一层外皮是4米多厚的石墙，内侧是一圈木构围屋带外环廊，二层外皮变为半米多厚的夯土墙，墙内是2米宽的木构环型马道，主要用于对外防卫，再向内是与一层木构围屋一体的二层木围屋，内侧亦有环廊。两层环廊随地势由前向后逐层升高，两层灰瓦屋檐和敞窗随之产生韵律感，特别是在后部转弯处，一边层层叠叠地升高，一边呈圆弧状转弯，景象奇丽，让人想起杜牧《阿房宫赋》中"复道行空，不霁何虹，高低冥迷，不知西东"等名句。

安贞堡前部　The front view of Anzhenbu.

堡内的后天井　The rear patio in the castle.

后部环道造型　The configuration of ring pathways.

堡内外围二层的环道，利于防御
The ring pathway on the top level of external wall is good for dense.

角部的枪炮孔
The firing holes for guns and cannons in the corner.

3. Anzhenbu Castle in Yong'an City

The famous Anzhenbu Castle is located outside Huainan town, similar in construction and aesthetic beauty to Shuangyuanbu Castle in Shuimei village, it is also said to have been constructed based on a complete set of design drawings. It was built for the Chi family during the reign of Emperor Tongzhi, located in the back of a cove in the mountain. It rises high with the mountain, overlooking the valley, standing tall in all its majesty and glory. In appearance it differs from the aforementioned castles in two ways. First, the two corner towers on either side of the front gates have roofs that are spired, four-cornered pavilion roofs. Second, the watchtower in the back of the castle has an additional attic sticking out of it. These two features of Sijiaolou buildings are similar to those funerary objects found in the tombs of the Han.

Overall, in the middle of the castle are enclosed houses, in a complicated but orderly fashion way. However, Anzhenbu Castle's most exquisite feature is its periphery. It is made up of two roofed causeways of different width stacked one atop the other, leading stepwise upwards towards the back of the castle. The tiled roofs and portholes following the causeways upwards provide rhythm, especially in the back where the layers are rising and turning simultaneously, turning the scenery unique and reminiscent of Du Mu's famous sentences in *On the E Pang Palace* such as: "The roofed causeway is stretched through the air; how can the rainbow appear without a clear sky? Both high and low are shrouded in mist, it is hard to distinguish east from west."

安贞堡侧后面　　The side and rear view of Anzhenbu.

外围的屋顶
The roof of the external ring.

堡内的装饰
The decorations inside the castle.

三、闽西南方土楼和府第式土楼

最标准的方土楼可以被理解为由四合院整体升高形成，或者是没有角楼的四角楼或碉楼。还有一些方土楼是综合形的，它们以不同的方式将院落式建筑局部加高，这样即有防御性，也维持了礼制建筑的仪态。方土楼虽然总是和圆土楼在一起，但我们认为它们有着各自的起源，方土楼的原型是中原井田制。

1. 永定县抚市镇实善楼

抚市镇是一座由几十座大大小小的、密集分布的方土楼形成的城镇，一座圆土楼也没有，这样的城镇在世界上恐怕也是独一无二的。这些大土楼也是由来自烟草业的财富在清代后期建造，当地的黄、赖两姓氏族建这些土楼有防御氏族之间械斗的目的，现在镇中的几座土楼废墟就是械斗的后果。

在众多大土楼间，实善楼不起眼，但它因在方土楼前有一座正面上有并排两个大门的院落而显得独特，这是因为有赖姓两兄弟要合建一楼又想有各自的入口和中堂形成的。两个大门内都是一侧为敞廊，一侧为共用花隔墙的天井，中堂也是只有一侧有堂屋间。中堂后，二进天井就合二为一，再后是4层高的土楼，土楼天井的中心有一口水井。这样的布局使土楼仿佛是院落后面的碉楼，有危险时，兄弟一起躲进土楼，平日则在相对舒适的院落中生活。

福建永定县洪坑村
Hongkeng village in Yongding county, Fujian.

抚市镇的土楼群
The earth building group in Fushi town.

Section 3 Earth Buildings in Southwestern Fujian and the Mansion-Style Earth Buildings

The most standard square earth buildings can be understood as a courtyard house stretched taller or a Sijiaolou without the corner tower or Diaolou. There are some fusion square earth buildings are combinations of different shapes and styles. Some are courtyard houses that are built higher in certain portions while other portions are not, providing defensive aspects while maintaining the ritual traditional courtyard architecture. Square earth buildings and circular earth buildings are often seen together, but we think that they have their own origins, of which square earth buildings being the Well-field system of the Central Plains.

1. Shishanlou, Fushi Town, Yongding County

Fushi is a town composed of densely and all different sizes square earth buildings. With no a single round earthen building to be found, this kind of town could very possibly be one of a kind. These earth buildings are products of the wealth of the tobacco industry, built in the late Qing Dynasty. The Huang and Lai families built these earth buildings due to conflict between the two clans, the ruins of the once buildings now lie in the town as a result of these struggles.

Compared to many other large earth buildings, Shishanlou does not seem like much, but its two front doors in the front of the building make it a unique. This is because there once were two Lai brothers who wanted to jointly build a building while still having their own entrances

中间的小方土楼为实善楼
The small rectangular earth building in the middle is Shishanlou.

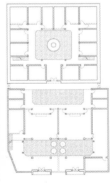

实善楼平面图
The plan of Shishanlou.

2. 五福楼

名为"五福楼"的一座大土楼是综合形方土楼的一种典型形式,院落建筑与土楼因素的结合更为整体性。这种楼的总体格局是一个礼制性很强的府第式大院落,这种院落当地称为"堂横屋",变成土楼后称"五凤楼",堂横屋的两侧横屋从前至后分段变为2~3层楼,形成前低后高的阶梯式,正堂则变成3~5层高的土楼。院落从整体看上去是一座土楼,也像是一座中间有大碉楼的古堡。

五福楼不是最标准的五凤楼,因为它不完全对称,但这样反而使它更有古堡的美感。

俯瞰五福楼正面　　The front perspective view of Wufulou.

平视五福楼前部　　The front level view of Wufulou.

and center halls. Just inside the front doors is a courtyard bisected by a wall, and behind the center halls is another courtyard with a well in the middle and is not separated into two halves. Behind this courtyard is a four-story earth building where the brothers stay in times of combat. In times of peace, they live in the relative luxury of the courtyard below.

2. Wufulou Building

Wufulou is a more basic form of fusion square earth building. The overall pattern of this building is traditional of highly mansion-style courtyards, this pattern became locally known as "Tanghengwu," and after it was turned into an earthen building, "Wufenglou Building." The houses on either side in Tanghengwu rise in segments to two, and then three stories tall, forming a staircase shape going towards the back of the earth fort. The main building in the center of the courtyard rises from three to five stories. The overall shape of Tanghengwu is that of an earthen building, but it also resembles a castle with a Diaolou in the middle.

Wufulou Building is not the typical Wufenglou Building, because it is not exactly symmetrical, but this also adds to its castle-like beauty.

五福楼内的石刻
The carved Stone of Wufulou.

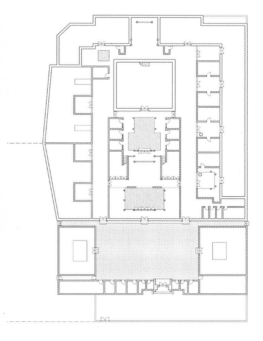

五福楼平面图　　The plan of Wufulou.

3. 永隆昌楼

永隆昌楼由两座方土楼组成，形式接近于五凤楼的叫福盛楼，接近方土楼的称福善楼，福盛楼为老楼，建造期早于福善楼，它的主体比一般的五凤楼整体感更强，更向城堡的意象偏重，两侧横屋由若干个带小天井的居住单元拼成，并与4层高的高大后围连成整体，且该外围的屋顶不采用歇山顶形式，故使福盛楼的构造看上去像是在一座五凤楼外面又包了大半个方土楼。

福善楼斜向地接在福盛楼背后，前院被福盛楼的后围挤去一角，楼主体完全是一座方土楼。

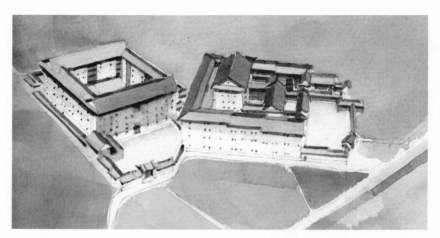

永隆昌楼鸟瞰图　The bird's eye view of Yonglongchang Buildiing.

永隆昌楼侧面　The side view of Yonglongchang Buildiing.

3. Yonglongchang Building

Yonglongchang Building is made up of two earth buildings, one similar to a Wufenglou Building called Fushenglou, and the other similar to square earth building called Fushanlou. Fushenglou, the older, being built before Fushanlou, has a main body that is more integrated than that of Wufenglou, more resembling the image of a castle. The houses on either side are apartments each with small patios but still connected to the four-story tall back wall. The roof of the periphery is not gable and hip roofs, which makes it look like the greater half of a square earth building was built around a Wufenglou Building.

Fushanlou attaches to the back of Fushenglou, with a corner of its front yard of which is taken up by Fushenglou. On the whole, Fushanlou is a square earth building.

福善楼中的装饰　The decoration of Fushanlou.

福善楼正面　The front view of Fushanlou.

福盛楼的主楼
The main building of Fushenglou.

4. 洪坑村福裕楼

在前述的洪坑村中，与振成楼隔河相对的福裕楼是振成楼建造者的长辈所建，建成期约在 1883 年，福裕楼的整体构造基本上是五凤楼形式，只是因其前楼比较高，后楼不仅更高而且通宽，并与两侧横楼相连形成方围，故增加了古堡的形态特征。与振成楼一样，楼的设计也是精巧别致，楼内以各种敞厅将本为堂横屋式的天街分隔成一个个小天井，天井的室外空间、敞厅的半室外空间和楼内的室内空间之间自如贯通，开合有序，空间极为丰富和舒适。由于空间层次多变，又是相互可望可至的楼房，楼似乎也因此变大了一些。

福裕楼内部　The inside of Fuyulou.

福裕楼的庭院　The courtyard of Fuyulou.

洪坑村村口　The entrance of Hongkeng village.

俯瞰福裕楼正面　　The front perspective view of Fuyulou.

4. Fuyulou, Hongkeng Village

In the aforementioned Hongkeng village, across the river from Zhenchenglou is Fuyulou, it was built by the ancestors of the builders of Zhenchenglou, around the year 1883. Fuyulou is essentially a Wufenglou Building only because its front building is high. Its back building is even taller and wider, connecting to the side buildings forming a square perimeter, most resembling a castle. Same as Zhenchenglou, the design is also exquisite, with verandas as great halls and open-aired "streets" throughout; the houses are separated into individual courtyards. The spaces of the outdoor patio, the semi-outdoor great halls, and the indoor rooms are freely connected and open and close systematically. The space is extremely sumptuous and comfortable. As the space changes and each room and position in the building are easily accessible, both visually and physically, the building becomes a little bigger.

上2图：洪坑村中另一座著名方土楼奎聚楼也为林氏家族所建
Another well known rectangular earth building in Hongkeng village was also built by Chen family.

四、广东大围楼

1. 大埔县泰安楼

大埔县位于饶平县北,县城位于湖寮镇,"寮"为木架草顶的简易棚屋,这让人想起客家人曾被自认为"文明人"的人指为棚民,被欺负的人中又有少数人指畲民才是棚民。

镇上的泰安楼是一座有石头外围的古堡式大围楼,楼主姓蓝,不是所有姓蓝的人都是畲族,然蓝姓确是畲族最主要的大姓之一。在早年,蓝式楼主自称客家人,近年又明确自己是畲族,祖先七百多年前从福建龙海迁来,楼建于清代乾隆三年(1738年),有客家民居特征,楼的后边呈弧形,有些像闽中土堡,也像客家围龙屋。

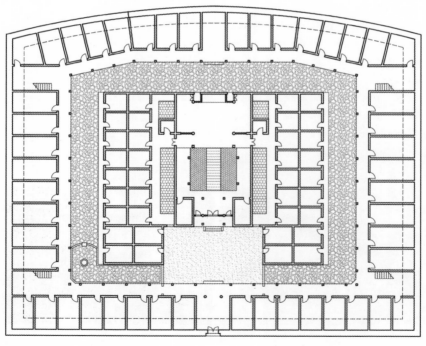

泰安楼平面图　The plan of Tai'anlou.

Section 4 Guangdong Grand Enclosed House

1. Tai'anlou, Dapu County

Dapu county, north of Raoping county, the maintown ship of which is Huliao town, "Liao" meaning simple shelters of wooden frames and thatched roofs. This is reminiscent of self-proclaimed "civilized" people referring to the Hakkas as shed people, a small number of which called She people shed people.

The town's Tai'anlou is a castle-like enclosure house with a stone outer wall, owned by a Mr. Lan. Not all Lans are She people, but the last name Lan is indeed one of the more common last names of the She people. In the early days, Mr. Lan claimed he was of the Hakka people, but more recently, he has insisted he is of the She people and that his ancestors moved to Huliao town from Longhai, Fujian 700 years ago. The building was built during the Qing Dynasty in the third year of rule of Qianlong (1738). It has Hakka characteristics, such as a curved back to the building, but also similar to the earth castles in middle Fujian Province and Hakka Enclosure Weilong House.

泰安楼大门　The gate of Tai'anlou.

楼内3层利于防御的外围环道
The third level ring pathway for defense.

泰安楼内部　The inside of Tai'anlou.

2. 东源县康禾镇仙坑村

东源县位于东江中游,是综合形古堡式大围楼最集中的地区,康禾镇仙坑村有一座八角楼,一座四角楼,八角楼最接近中国传统形式的古堡,它的整体构造是内外两圈,内圈为一座大型四角楼,外圈则只有一道城墙,四角有角楼,内外共8个角楼,所以俗称八角楼。

楼前面在横屋部位均有突出物,结合水塘前的照墙形成前院,这样在满足风水学要求的同时也有一定的防御性。外围由粗大的条石砌成,厚达1米,加上上部的垛墙高达6米,角楼顶则高达10米。

楼内围建于清代乾隆年间,屋主叶氏先祖在外做官多年,回乡后耗银数万两建成此楼,楼在很长时间内并没有外围,据传是道光年间太平天国运动爆发后才加建。

四角楼的建造年代晚于八角楼,防御性比八角楼弱,但它更大,构造也更复杂,更考究,内部装饰也更好。

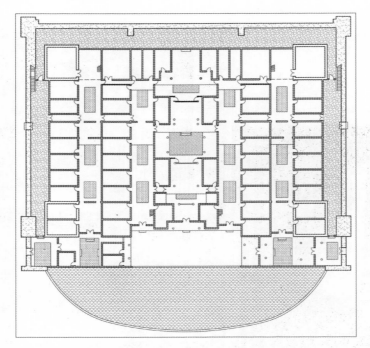

仙坑八角楼平面图　The plan of octagonal building in Xiankeng.

2. Xiankeng Village, Kanghe Town, Dongyuan County

Dongyuan lies in the middle reaches of the Dongjiang River, and is the area with the highest concentration of comprehensive castles. In Kanghe town, Xiankeng village are an eight-corner-tower building and a four-corner-tower building. The eight-corner-tower building closely resembles traditional Chinese castles. Its general structure is two concentric layers where the inner is a four-corner-tower building, said Sijiaolou, and the outer being a rectangular wall with a corner tower on each corner, making a total of eight towers, thus named eight-corner-tower building.

From the houses on either side are protrusions that connect to the screen wall separating it from the pond, forming a front yard; this way, while satisfying feng shui teachings, the building also have some aspects of defense. The outer wall is made of one-meter thick coarse stone with buttresses six meters tall and towers as tall as ten meters.

The inner portion was built during the reign of Emperor Qianlong. The original owners, named Ye, worked for the government far from home, spent countless of tael of silver on the construction of this building. For a long time, the building did not have an outer wall, and it is said that it wasn't until after the Taiping Rebellion during the reign of Emperor Daoguang that the outer wall was constructed.

The four-corner-tower building was built after the eight-corner-tower building. It is weaker than the eight-corner-tower building defensively, but it is bigger, more complex and sophisticated, with better interiors.

堡门的石头门闩
Stone latch of the castle gate.

外围城墙上的马道
The bridle way on the external ring wall.

八角楼内两道城墙之间　The view between two walls in the octagonal building.

3. 乐村石楼

乐村位于蓝口镇东部，村中心是张氏家族的大石楼，楼从乾隆年代创建，嘉庆年代竣工，这座"外可御敌内可安居"的石楼，在我们见过的东江古堡中是文化融合性、建筑综合性最强的。

石楼的壮观首先来自于它大的风水格局，这种格局赋予了石楼所属的宇宙，石楼成为这个小宇宙的中心，不论站在楼后的镇山上还是楼对面的案山上向下望，石楼体量与山谷尺度相融相衬，巨大的石楼前真仿佛可以藏风纳气。

石楼正屋两侧有高大的外围围护，外围前后各有角楼，正屋后部是隆起的化胎，弧型围墙围护，化胎左角随弧型围墙加建一座格外高耸的弧形炮楼，加上前部增建的两座塔楼现在只剩一座，故石楼现状共有 6 座塔楼，中心建筑呈方形，加上前水塘后化胎，整体平面呈不规则椭圆形。

"化胎"是风水学反映在建筑上的一种元素，按风水学的一般要求，房屋背靠镇山以接"龙脉"，为防龙脉上水土流失，要种"风水林"，而树木又不可种

俯瞰石楼正面
The front perspective view of Stone Building.

石楼后部的化胎和炮楼
The *Huatai* and cannon tower at the rear of the Stone Building.

3. Stone Building, Lecun Village

Lecun village is located in the eastern part of Lankou town, the village center is a large stone building belonging to the Zhang family. The building's construction began during the reign of Emperor Qianlong and was completed during the reign of Emperor Jiaqing. This stone building, that can protect on the outside and be a comfortable home on the inside, is the most culturally comprehensive and architecturally diverse of the castles around the Dongjiang River.

Stone Building is spectacular in its large *feng shui* layout, which makes the area into its own universe with Stone Building in the middle of it. Whether one is looking on it from the Zhenshan, the mountain behind the building, or the anshan, the mountain in front of the building, Stone Building's scale blends in well with the surrounding valley. The space in front of the building is capable of storing wind and accepting air.

On either side of the main building are side buildings with outer walls behind them, with a tower on each corner. Behind the main building bulges a *Huatai* enclosed by a semi-circular outer wall. Along this wall is a block house that rises high, and that combined with two

得离屋太近，恐树根破坏后墙地基，风水师便在屋后创造出个化胎概念来，龙脉在化胎中孕育，然后向前接屋后的"穴门"，再接至祖堂的祖宗神位上，天地人神便是一体了。化胎如此重要，便需要围护，可以用高墙，如乐村石楼的做法；也可以用围屋，围屋随弧形的化胎也必呈弧形，弧形适宜山坡地，外凸弧形也利于屋后分水。

如果维护化胎的是围屋，所谓"围龙屋"就形成了，它是客家民居的一种重要形式，与圆土楼一样，这种形式不大可能是客家人从中原带到南方的，它的来源可能与圆土楼类似。

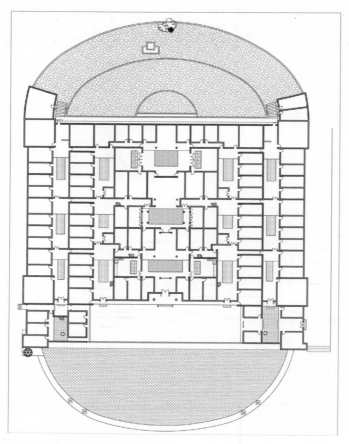

石楼平面图　The plan of Stone Building.

additional towers built in front of the building, of which only one remains, brings the total number of towers to six. The main body of the building is rectangular, but with the pond in the front and *Huatai* in the back, the entire structure is an irregular oval.

Huatai is an architectural element in *feng shui*, which requires the house to have its back to a mountain and connect to the "dragon vein." Dragon vein are mountain ranges or rivers that are continuous over long distances, and in order to prevent the loss of water and soil erosion of dragon vein, feng shui forests are planted to keep the soil in place, but not too close to the house because tree roots are capable of doing irreparable damage. From this sprang the idea of the *Huatai*, which carries the essence and function of dragon vein. The "Xuemen" connecting to the back of the house, and also connects to the ancestors' tablet in the ancestor hall, heaven, earth, man, and god all become connected. *Huatai* is important, and therefore need to be protected by walls, such as in Stone Building, or using enclosure house. Enclosure house, following the *Huatai*, also would become semi-circular, which would also need to be suitable to the hillside and helps water that flows down from the mountain around the structure.

If enclosure houses are used, "Enclosure Weilong Houses" are formed, which is an important style of the Hakka people. Its origins are similar to round earthen buildings and not likely to be brought south from the Central Plains.

炮楼上的枪炮孔　　The holes for firing guns and cannons in the tower.

角楼上的枪炮孔外面
The exterior of the holes of guns and cannons in the tower.

前部加建的角楼
The additional constructed corner building in the front section.

化胎全貌
The overview of *Huatai*.

穴门处的五行风水石
Five elements of *feng-shui* stone at the temple door.

4. 三顺楼

三顺楼位于义合镇段的东江边，其整体上是一座巨大的环有围墙的古堡，外轮廓呈不规则长圆形，其内部建筑主要由 6 个部分组成，位于正面东南向右侧的大体量建筑为该古堡的主体建筑，呈方形，其外表像一座围楼，前面左右两端有角楼，后部居中为高大的望楼，其大门也是整个城寨的正门。

整个城寨的第 2、第 3 个组成部分是位于第一部分后面的两座独立高楼，都是斜向摆放，较正向的一座目前完全独立，斜向的一座附带一些附属房舍，两座楼之间的城墙上有朝向东北方向的堡门。

左侧的堡门　　Castle gate on the left.

大门侧面的枪炮口，对准大门前
The gun/cannon firing hole at the side of the door, aiming the front yard.

右侧的堡门　　Castle gate on the right.

第4部分位于主体围楼西北面,是一座面向东北面的一进小院,后部有两座角楼,凸出的一座上有朝向西南面的门。

第5部分是一座祠堂,位于第4部分南面。

第6部分为3座散落的附属建筑,都位于1、4、5部分之间,而它们之间还夹着一个位于正门左侧不远的小堡门。另古堡西北角还有一个较完好的堡门,门楼高大。

据楼主人讲,他们祖上是在元朝时从江西抚州迁来的,此大楼已建成两百多年了,祖上靠卖炭经商致富,清朝时曾帮河源官府拒匪,得过朝廷封赏。三顺楼古堡整个占地面积约8000平方米,加上相邻的一座大屋,这组建筑群中各种楼如望楼、角楼、大门楼等共有10座,使其轮廓非常壮观。

三顺楼主体正面　The front view of Sanshunlou's main buildings.

4. Sanshunlou Castle

Sanshunlou Castle is located in a town on the Dongjiang River called Yihe town. It is an oversized castle surrounded by an outer wall, its contour an irregular ellipse. Its internal structure consists of six parts, with the building in the southeast corner being the main architecture. The main architecture resembles a enclosed house; it has two towers in the front and a huge watching tower in the center of the back wall. The front door to this building is also the main gate to the whole castle.

Behind the main architecture are two independent towers, the second of which is placed at an angle from the first. The first building, square with the main building, is completely independent while the second has attachments to it. The wall between these two towers has an auxiliary northern gate.

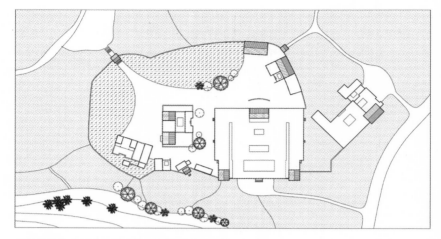

三顺楼整体平面图　The plan of all buildings of Sanshunlou.

楼内的堡院　The courtyard in Sanshunlou.

In the southwest of the main building is a courtyard facing northeast with two towers in the rear. One of the towers sticks out from the building and also has a door leading southwest. An ancestral temple lies south of the courtyard. Three outbuildings surrounded by the main architecture, the temple, and the courtyard, are located next to a small gate west of the main gate. A large, relatively intact gatehouse opens to the west.

The inhabitants' ancestors migrated there from Fuzhou, Jiangxi Province during the Yuan Dynasty. This building had stood for over two hundred years, the ancestors of which became wealthy by selling charcoal. During the Qing Dynasty, they helped the Heyuan official fight off bandits, which earned them a reward from the court. Sanshunlou Castle covers an area of about 8000 square meters, including a large architecture; this group of buildings vary in types, including watch towers, corner towers, and gates, creating a spectacular outline.

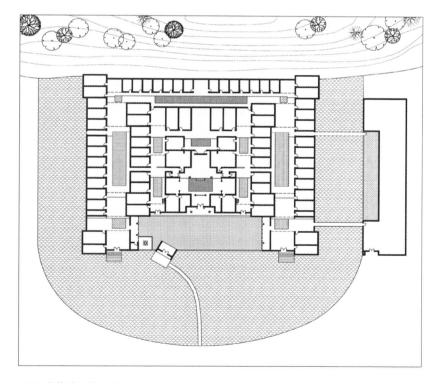

山下八角楼平面图　　The plan of eight-corner-tower of Shanxia village.

243

5.叶潭镇山下八角楼

古堡的形象标志除了高大的城墙以外,便是碉楼高塔,竖直方向的丰富变化对于建筑造型显然是重要的,特别是心灵属性强的建筑或建筑群,有高塔的教堂、清真寺、佛寺等与高塔的同类结构给人的视觉特别是心理上的感觉就是不一样。而中国的传统建筑发展到明清,愈发地强调水平方向的复杂构造,不重视向上发展,传统民居中就更缺乏高耸物。然而,山下八角楼这座并不特别大的客家围屋竟耸立着8座大炮楼,以至让人感觉它是以炮楼为主的,加上它三面临水,通向大门的路是一条弯曲的石桥,故让人第一眼看到它时,不仅在时间上有恍惚感,在空间上也有恍惚感,觉得自己是否站在了欧洲的一座古堡前。

由石础、石梁、石板搭成的曲桥从水塘前部的中心出发,弯曲着通向八角楼正面斜向布局的门楼的前台,进门分辨,这座大楼的格局是一个大四角楼套着一个小四角楼,内部的小四角楼为一座整体型围楼,外围四角楼没有前围,以一道既似门又似廊庑的构造与内环相连,正中部分是连接着斜向门楼的近4米高的高墙,结合水塘形成防御性。

八角楼正面
The front view of the Castle.

楼内的匾额
The inscribed board in Castle.

5. Eight-corner-tower Castle in Shanxia Village, Yietan Town

The image of the castle, besides tall walls, centers on lofty towers, which are clearly architecturally important, especially for spiritual venues. The feeling that one gets from churches, mosques, temples, and other such structures that have towers and those without them is vastly different. Traditional Chinese architecture development until the Ming and Qing dynasties emphasized ground-level development and complexities, and did not focus on the third dimension, so traditional vernacular dwellings are even less likely to contain tall structures. However, eight-corner-tower Castle in Shanxia village, a Hakka building of no significant size, has eight towers, which gives one a sense that it is mainly a blockhouse. When one sees this building, surrounded by water on three sides with a curved stone path leading to the front gate, one is transported in time and space to a European castle.

The completely stone bridge, from foundation to beams, arches over the center edge of the water reservoir and curves to meet the platform before the diagonal-facing front gate. The building is made up of two concentric four-corner-tower buildings connected by doorways that, at a distance, look like a long veranda. The inner is a enclosed house while the outer has a four-meter high wall instead of a row of rooms in the front, which in addition to the water reservoir, makes the building much more secure.

八角楼侧面
The side view of castle.

俯瞰八角楼
The perspective view of the eight-corner-tower Castle.

245

6. 紫金县德先楼

客家古堡虽多，但真正属于山地建筑的并不多，德先楼可以算一座。它所在的南岭镇位于群山之间的一个小盆地，德先楼建在盆地边缘一座陡峭高山的高坡上，面向宽阔盆地，左右两条溪水在楼前汇合流向北方，这是为了获得好的风水形势。

根据德先楼主人保存的资料，该楼始建于清代光绪20年，但建筑不断被山洪造成的山坡滑坡毁坏，后来挖深沟疏浚山洪，并以风水林、化胎等设施巩固山坡，再将整体大屋分建在两层台地上，才形成现在占地达3000多平方米，建筑面积5000平方米，有6座高大角楼，与山体融成一体的恢宏建筑。

大楼三面为溪水环抱，只在正面和左侧有两座小拱桥沟通内外，两层台地的高差约7米，前端的两个角楼内有楼梯，竖向连接上下，低台地上的建筑部分除角楼外高二至四层，底层主要是仓库、畜圈、粮食加工间等，有少量独立对外的出入口，高层台地上的建筑是主要人居部分，另有坡道直接相通，以此实现了人物分流。

人流主入口在高台地左侧，进门后是片宽阔的平台，一面是主楼正面，一面是矮护墙，护墙下有下层台地上建筑中的一个天井，对面有分别为三层和七层的角楼，空间层次丰富，景象壮丽。

德先楼主人钟雨金先生绘制的中国画
A Chinese painting by Mr. Zhong Yujin, the owner of Dexian Castle.

德先楼正面　The front view of Dexian Castle.

德先楼的前庭　The front courtyard of Dexian Castle.

6. Dexian Castle, Zijin County

Although Hakka castles are numerous, those built on mountains are not, and Dexian Castle is one of the few. In Nanling town, located in a small basin between mountains, Dexian Castle was built on a steep slope on the edge of the wide basin. On either side of the Castle are two streams that converge before it before flowing away to the north, giving the area better feng shui.

According to the information saved by the landlord, the building was built in the Qing Dynasty, in the twentieth year of the reign of Emperor Guangxu. However, the buildings were under continuous abuse from flash floods caused by the terrain. Later, deep trenches were dug to counter these floods, *feng shui* forests planted and Huatai installed to consolidate the hillside. The entire Castle was built on two terrace in the mountain face, covering a lot of 3000 square meters and a built-up area of 5000 square meters. All this, including six towers, melts into the mountain into a magnificent building.

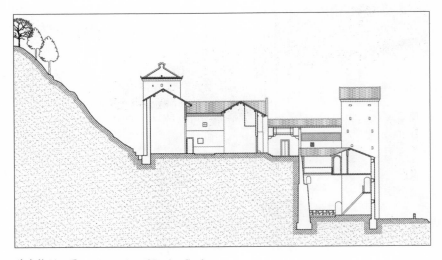

德先楼剖面图　The section of Dexian Castle.

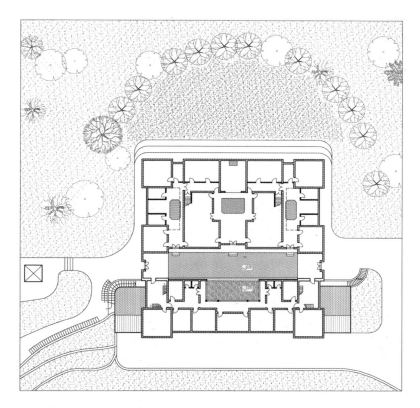

德先楼平面图　The plan of Dexian Castle.

Surrounded by streams on three sides, two small arch bridges, in the front and on the left, connect the castle's interiors and the outside world. Staircases in the two forward towers connect the steps, differing in height by seven meters. The buildings on the lower step, besides towers, are two to four stories high. The bottom floor mainly consists of warehouses, livestock, and mills, and a few exits directly out of the castle. The higher step is composed primarily of housing and is connected to other parts of the castle by ramps, separating man from beast.

The main entrance is on the left of the top terrace. Upon entering through the door, one sees a wide platform with the front wall of the main building on the left and a low parapet wall, under which is a patio of the lower terrace. Looking out over the parapet, a three-story and a seven-story tower create a space with depth and a magnificent sight.

楼内正堂
The main hall.

楼内前部底层的回廊
The corridor of the ground level in the front.

7. 新丰县儒林第

儒林第位于新丰县城西南的梅坑镇大岭村,大围楼的主入口和最佳观赏面都在左侧,从这里可见由门楼、角楼、望楼构成的美丽构图。门楼正面和门内通道都是斜向,走在门道里可以看到这段斜门将两圈围楼连接起来,人可由此进入两圈围屋间的环道,也可再过一段门道穿过横屋直接走入内圈围屋的前天街。

外围上除有两座门楼外还有四座角楼、一座望楼,加上内围前面的两座角楼,全围共有9座楼。巨大的望楼平面呈长方形,四角外凸,高4层,局部5层,自身就像一个小四角楼。望楼所在的外围段落拖后角楼一段,两外角抹圆,于是产生了些围龙的迹象。总之,儒林第整体构造虽然规整,就是内外两圈围,但其局部变化无穷。

儒林第侧面　The side view of Rulindi.

7. Rulindi, Xinfeng County

Rulindi (the scholar's mansion) is located in Daling village in Meikeng town in the southwest of Xinfeng county. The main entrance and the best view are both on the left. The view is a beautiful composition of the gatehouse, turrets, and watchtower. The door and corridor of the gatehouse are both positioned at an angle and connect the inner and outer rings. From this corridor, one can enter the walkway between the two rings or pass through another door to the open-air corridor within the inner ring.

Besides the two gatehouses on the outer ring, there are also four turrets and one watchtower, plus two turrets in the inner ring, making a total of nine towers. The watchtower is rectangular with corners that protrude from the faces. The tower is four stories high, although some areas have an extra story. In general, it is very similar to a four-corner building. The side of the watchtower is further out than the two turrets on the same side. The rounded corners on this side create a reminder of Weilong. Although the general structure of Rulindi is organized, two concentric rings, but the structure varies greatly within.

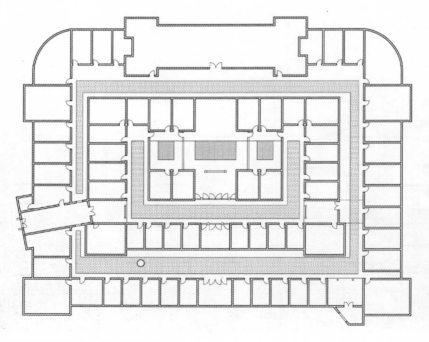

儒林第平面图　The plan of Rulindi.

8. 增城市邓村石屋

增城市派潭镇南部的邓村中有座灰砖大屋，屋主为石姓客家人，大屋整体呈围屋型，建筑总宽80多米，深40多米。前有水塘和照墙围起前坪，屋后无围龙屋但有化胎。

大屋最醒目的要素是屋内左部立有一栋长方形、三开间、7层高的大碉楼，比粤北的碉楼高一些、细一些，但比山西皇城相府的河山楼粗大，像是将义合镇三顺楼的望楼加高，但它在建筑群中所处的位置像河山楼，有它的存在，这座大围屋就有古堡性质了。

8. Stone Castle, Dengcun Village, Zengcheng City

In Dengcun village, in southern Tanzhen in Zengcheng City, is a grey stone house owned by a Hakka family which surname is Shi. The general organization of the house is a quadrangle of eighty meters wide by forty meters long. A reservoir and a screen wall enclose the front courtyard. The back of the house does not have the Enclosure Weilong House, but has a *Huatai*.

The most striking element of the house is a rectangular, three-intercolum, seven-story tower on the left. This towers is taller and thinner than most towers in northern Guangdong, but bulkier than Heshan Tower in the Huangcheng Xiangfu in Shanxi, sort of like the Yihe town watchtower, only taller. However, the house's location is similar to that of Heshan Tower. With the tower, this house becomes more castle-like.

邓村石屋全貌　The overview of Stone Castle in Dengcun village.

大屋内的院落
The courtyard inside the house.

大碉楼正面
The front view of the Diaolou.

9. 深圳龙岗鹤湖新居和龙田世居

深圳境内的客家大围屋体量更为巨大，城墙和炮楼凸显出它们的古堡性质，龙岗街道的"鹤湖新居"是目前所发现的面积最大的客家围屋式古堡，其面宽166米，深104米，总占地两万多平方米，建筑面积1万5千平方米。其整体由4个围组成，中心一个方形围，左右两边各一个外侧斜向的边围，后部一个扁长的后围，这种格局表明该大屋应是逐步形成现在规模的。大围整体前排4个炮楼，中心方围后角两个炮楼，中间一座望楼，外、后围后角两炮楼，中间还有一望楼，全围共十座高楼，可惜有数座已坍塌。

大围屋于清代嘉庆年间建成内围，围正面有三个出入口，都开在面宽90多米的内围上。

龙田世居规模虽小，但它构造严谨，方形主围左右两侧还有两道由众多小四合院排列而成的横屋，使内外居住空间都是单元构造。再外是一圈宽阔的护城河，入口也是由桥与外相连，与山下八角楼相似，它虽无八楼，也有五楼，即四角楼外还有后排居中一望楼，后部无围龙，但有半圆形后花园。

鹤湖新居正面
The front view of Hehu Xinju.

大围内后部的望楼
The watchtower in the rear of the enclosure house.

9. Hehu Xinju and Longtian Shiju, Longgang, Shenzhen

Within Shenzhen, Hakka houses are even greater in number; walls and towers highlight their castle nature. In Longgang, Hehu Xinju currently resides the largest Hakka castle, measuring in 166 meters wide and 104 meters with a total area of over 20,000 square meters and a built-up area of over 10,005 square meters. The overall structure is composed of four quads; the center is the main quad. On either side of the main quad are two long, oblique quads. Behind these three quads is a wide and narrow back quad. This arrangement suggests that the house was built in installments rather than all at once. The entire has ten towers, four block towers on the front wall, two block towers on the back wall of the main quad, two block towers on the back wall of the back quad, and one watchtower on the back wall on both the main quad and the back quad. Unfortunately, some have collapsed over time.

The main quad was built during the reign of Emperor Jiaqing of the Qing Dynasty, with three gates spread over the ninety-meter-wide front wall.

Although the scale of Longtian Shiju is small, its structure is strict. There are many row houses consisting of small Siheyuan on both sides of the rectangular main building, thus the residential place of interior and exterior are all modular houses.

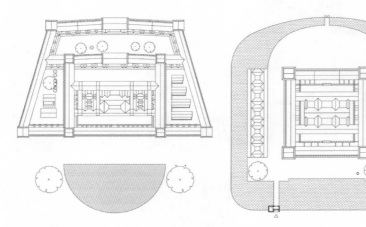

鹤湖新居平面图
The plan of Hehu Xinju.

龙田世居平面图
The plan of Longtian Shiju.

龙田世居的护城河　The moat around Longtian Shiju.

10. 深圳东莞碉楼

除了大围楼，深圳地区另一种古堡形式是村落中的高碉楼，碉楼的形态类似皇城相府的河山楼，由于华侨因素，它们都有一些南洋式的装饰，但这种碉楼形式应该不是来自南洋。如果说大围楼强调的是设施防御的话，那么这种村落强调的是信息防御，高高的碉楼可尽早发现敌情，当然，它也可以作为防御设施。

这种高碉楼最密集的地方在东莞市凤岗镇黄洞村至深圳市观澜镇油村一带，不大的区域内集中了十几座高碉，点缀在平淡的排屋间。高碉的平面尺寸多为五六米见方，比常见的围屋上角楼还小1米，而其多为六七层高，最高者八层，故显得细高。当地人讲，从前这片区域中有上百座高碉，想当年这里的景象恐怕比如今有上百座新高楼还要壮观些。

2图：东莞凤岗镇的碉楼　　Two above: tall Diaolou in Fenggang town in Dongguan.

排屋内的彩塑
The colored sculpture.

观澜油村
Youcun village in Guanlan town.

油村碉楼的顶部装饰
The top decorations on the tall Diaolou in Youcun village.

惠阳地区的碉楼　Tall Diaolou in Huiyang region.

10. Diaolou in Dongguan, Shenzhen City

Besides large enclosed houses, Shenzhen also has another type of castles, tall Diaolou. The Diaolou is similar to the Huangcheng Xiangfu-Heshan Tower. The decoration of the Diaolou is Southeast Asian in origin due to migration to and from Southeast Asia, but the style of Diaolou itself is not from Southeast Asian. If large enclosed houses emphasized defense, these villages are more preventative in nature and the tall towers provide a means of detection before the tower comes under siege. They, of course, also have defensive faculties.

These tall structures are most concentrated between Huangdong village, Fenggang town, Dongguan City and Youcun village, Guanlan town, Shenzhen City. There, a dozen or so towers are located within a small area, decorating neat rows of houses. The cross section of these towers are square with widths of five to six meters, which is one meter narrower than the commonly seen towers of enclosed house. These six-story, seven-story, sometimes eight-story towers seem even thinner and taller. The locals say, hundreds of these Diaolou stood in this area long ago, a scene more spectacular than if hundreds of towers were newly built here today.

11. 开平碉楼

据记载，观澜镇油村的客家人在古代是从北方河北省先迁广东开平，清代中期才迁至观澜的，那时开平已有碉楼，不过形式还不是现在的华侨洋楼，而是粤北客家四角楼那种，加之清代中期时客家人迁往粤西的很多，所以，这种迁徙过程让人觉得，开平碉楼与深圳、东莞客家碉楼之间可能是存在特殊关系的。虽然现存的开平碉楼以多样的南洋建筑风格著称，但仔细看，碉楼的基本结构还是粤北碉楼的那几种形式。后期的碉楼使用了新的建筑材料——水泥（当地人称"红毛泥"），使建筑造型更为灵活。

深圳、东莞的碉楼平日大多不做居住用途，而开平的碉楼平面面积较大，楼内即可居住，这与此地在清代末年族群械斗和匪患问题比较严重有关。

众楼是多户人家共建的避难楼，多位于村落边，日常不住人
The Refuge tower is built by several families, normally located at the edge of the village for refuge usage, not for daily living.

更楼相当于烽火台
The watchtower is equivalent to the beacon tower.

11. Diaolou in Kaiping County

According to the records, the Hakka people of Youcun village, Guanlan town, first moved to Kaiping, Guangdong from Hebei Province, and moved to Guanlan during the mid-Qing Dynasty. Then, Kaiping had some Diaolou, but not in the form of Southeast Asian tower, but instead in the style of eastern Guangdong, Hakka four-corner-tower architecture, Sijiaolou. In addition, many Hakkas moved to western Guangdong during the mid-Qing Dynasty, which leads people to think that there is special relationship between Kaiping Diaolou and Dongguan Diaolou of Shenzhen. Although the existing Kaiping Diaolou are known for a variety of Southeast Asian architectural styles, a closer look at the basic structure of the Diaolou reveals the basic structure originating in northern Guangdong. Later, the Diaolou were built with a new material-cement (locally known as "red hair mud"), which is more versatile, allowing architectural modeling to be more flexible.

Diaolou in Dongguan and Shenzhen are not generally used as residences, but Kaiping Diaolou have bigger cross-sections, enough for people to live in, which was made necessary by conflict during the late-Qing Dynasty.

开平自力村中居楼和众楼的组合
The combination of Resident tower and Refuge tower in Zili village in Kaipi.

开平碉楼按功能可分为居楼、众楼、更楼3种，居楼既有防御性，日常也可居住，本页3图中的碉楼均体量巨大，做工精良，均为居楼

The Diaolou have three types: resident tower, Refuge tower and Watchtower. Resident tower has defensive function, also for daily living. On this page, the Resident towers all have huge volume, and very well crafted living quarters.

第六章 其他古堡
Chapter VI Other Castles

大金川峡谷中的古堡遗迹　The Ruins of castle in Dajinchuan River Valley.

虽然中国的古堡文化意识目前还比较淡漠，中国不是公认的世界古堡大国，但要将中国现存的古堡建筑信息较全面地收集起来，仍然是一项庞杂的工作。我们前面只是粗略地介绍了几个古堡集中区的情况，其他地区的古堡、其他类型的古堡，我们下面只能做一些更简略的介绍。

一、军事转民用的古堡

1. 江西省寻乌县羊角水古堡

在中游遍布大古堡的东江，其上游也是如此，东江源头的赣州市寻乌县大竹岭地域也是客家地区，那里也是长江水系和珠江水系的分水岭。分水岭北面有条小河名为湘水，湘水边有座筠门岭镇，镇旁有座羊角水古堡，从前是政府的军事戍堡，现在是客家围村。

四川丹巴县的藏民碉房　　Tibetan house in Danba country in Sichuan.

Castle cultural awareness in China is still relatively meager, so China is even less internationally recognized as a nation containing castles. Nevertheless, China has enough castles so that in order to compile detailed information on all of them is still a tall order. Previously, a few areas more concentrated with castles were introduced and discussed, but the following will be an even briefer introduction of other types of castles.

Section 1　Military-turned-civilian Castles

1. Yangjiaoshui Castle, Xunwu County, Jiangxi Province

The castle-abundant, middle reaches of the Dongjiang River flows from upriver where there are also many castles. It flows from Dazhuling region of Xunwu County, also a Hakka region, and also watershed of the Yangtze River and the Zhujiang River. North of the watershed is a river named Xiang River. On one bank of this river is Junmenling town, beside which is Yangjiaoshui Castle. This castle was formerly a government military garrison; now, it is a walled village of the Hakka people.

According to the records, during the reign of Emperor Chenghua in the Ming Dynasty, Yangjiaoshui Castle only housed a few hundred people, being a smaller garrison. Thousands

老城墙　Ancient city wall.

古堡大门　The gate of the castle.

据记载：明成化年间时，羊角水古堡还只是个屯兵百人的小军事戍堡，堡周边有上千户人家。当时盗匪横行，大股盗匪来时，戍堡内的明军只能闭门自守，外面的老百姓就遭了秧，每年几乎都要让盗匪洗劫一两次。为了摆脱这一窘境，百姓们希望能将戍堡扩大，他们与军队共同住在城堡里。在嘉靖二十三年（1544年），一座独特的军民共建共守的城堡出现在羊角水，当时的规模"周三千尺"，东南西北各有城门。

现存的羊角水古堡的周长近一公里，与"周三千尺"差不多，石磊的城门虽破旧，但仍巍然屹立，明代城墙也清晰可辨。它如今整体的格局是一个大型客家围村的外面又包了一圈城墙，城墙呈不规则圆形，外圈的房屋和道路都是环型的，向内逐渐过渡为方格网型，但网格不规则，目的是为了防御。

从临湘水的南门进入古堡，里面有个小城隍庙，这是一般这种规模的古堡和围村没有的，它显示这里是一座城。由于羊角水古堡扼水路要冲，太平时这里不可避免地会成为商埠。

堡内的节孝牌坊　The honorific arch for filial piety and chastity in the castle.

of people lived in the area surrounding the castle and were often victims to banditry. The soldiers garrisoned in the castle, small in number, could only close their doors and were unable to help the people. Each year, bandits looted the village once or twice. The villagers, not wishing to be left in the cold by the army that is helpless to defend them, wanted to expand the castle and live in the castle along with the soldiers housed there. During the 23rd year of the reign of Emperor Jiajing (1544), a unique castle, built by villagers and soldiers together, appeared in Yangjiaoshui, being "perimeter, three-thousand *chi*" with a gate in each direction, north, south, east and west.

Today, the perimeter of the castle measures at one kilometer, about "three-thousand *chi*", the stone gate, though old, still stands. The wall that was constructed during the Ming Dynasty is still discernible. The general scheme is a large Hakka courtyard with a roughly round wall surrounding it. The outer layer of houses and roads are all circular, slowly transitioning into an irregular grid pattern closer to the center courtyard. The irregularity became a defensive measure.

Entering the south gate of Yangjiaoshui Castle, one immediately happens upon a temple of the town's deity, something not found in walled villages and castles of this size. The presence of this temple indicates that this castle is also a city. Due to the Yangjiaoshui Castle's position at the watershed of two major waterways, during times of peace, this city inevitably becomes a commercial port.

上2图：改为民用之后，古堡内增加了许多宗祠建筑
Above: after changing to civilian use, many ancestral halls were built inside the castle.

2. 贵州省安顺市屯堡

安顺是汉族、布依族、苗族混居区,但有一批至今住在山间屯堡中的汉族人在习俗、服饰上与周边的布依族人、苗族人很接近,那是因为他们的祖先在明代初年就在此定居了,他们的文化慢慢融入了当地文化,但他们永远会记得,祖先是明朝的军户,来自江西、安徽等地,他们的屯堡原来是明军的戍堡。

云山屯堡位于一个山坳中,山口筑上城墙、城楼,易守难攻,城门里面实际上是一个小村。除城门一侧,环村都是陡峭的山峰,村里的房子不仅用石头垒墙,还有石板做屋顶,极具特色。

2图:屯堡中的石头碉楼
Two left: the stone towers in castle.

2. Official's Military Fortress, Anshun City, Guizhou Province

Anshun is a place where people of the Han, Buyi and Miao people live together, but groups of Han people who live in a fortress in the mountains have customs and clothing very similar to the nearby Buyi and Miao people. This is because their ancestors settled there in the early Ming Dynasty. The local culture became integrated into their own, but they still remember and honor their ancestors, who were military families from places such as Jiangxi and Anhui. Their village was once a fort for military garrisons.

The Yunshantun Fort is located in a cove; walls and towers were built at the opening of the cove to make the castle easily defensible. The inside castle is actually a small village. Besides the side of the front gate, the village is surrounded by steep mountainsides. The houses in the village not only have stonewalls, they also characteristically have roofs of stone as well.

云山屯城墙内部   The inside of the city wall of Yunshantun fotress.

云山屯的堡门
The castle gate of Yunshantun.

屯内中心部位的公共空间
The public space in the central section of Yunshantun.

屯内有堡院痕迹的民居
The houses can be traced back to previous walled courtyard in Yunshantun.

二、湖北省南漳县古堡

南漳县位于襄樊市南部,境内多山,明末曾经攻掠山西沁河的陕西农民军在败于明军和最终败于清军后都进入过这片山区,明清时期数次白莲教起义军也在此活动过,当时农民军和当地人都会建城堡保护自己。

南漳县的群山中现存几十座古堡遗址,古堡多为石头垒筑,只强调防御性,没什么建筑艺术的追求,但一些古堡建在奇山异水中,又规模宏大,所以能具有一种特殊的美。如春秋寨,它位于被一条名为茅坪河的河流3面环绕,周边都是悬崖峭壁的山脊上,古堡长近500米,石头寨墙里面有100多间石头房子。

立于陡峭山脊上的春秋寨　　Chunqiuzhai Fort stands on the steep ridge of a mountain.

Section 2   Nanzhang County Castles, Hubei Province

The mountainous Nanzhang county is located in south of Xiangfan City. During the late Ming Dynasty, the Shaanxi peasant, having fought and lost to the Ming, and later Qing, army in Qinhe, Shanxi, passed through these mountains. During the Ming and Qing dynasties, the White Lotus Rebellion held movements in this area as well. Locals and armies alike built castles while here to protect themselves.

In the mountains of Nanzhang county remain dozens of castle ruins, mostly made of stone. They were built with emphasis on defense, not of architecture or artistic value. Nevertheless, many castles with vast grounds were built on strange mountains or waters, giving them a particularly peculiar beauty. For example, Chunqiuzhai Fort is surrounded on three sides by Maoping River and cliffs and ridges of mountains, with length of almost 500 meters. Within the stonewall of the fort, there are more than one hundred stone houses.

春秋寨的堡门   The gate of Chunqiuzhai Fort.

三、羌藏地区的民间古堡

现在的羌族人主要居住在四川省的阿坝藏族羌族自治州中几条河谷里,他们的邻居嘉绒藏族等与古代羌人应该有族源关系,"嘉绒"就是河谷的意思。自古以来,那里民族纷争多,盗匪多,野兽也多,为了人和牲畜的安全,当地人的传统民居都是有一定防御性的多层碉房,下层用于关牲畜,上层用于住人和仓储。

更重要的防御性建筑是一种细高的碉楼,比皇城相府的河山楼和广东碉楼更高更细,形似烟囱,它被用来防御、仓储,也像意大利的圣吉米尼亚诺的碉楼一样,用于家族间的攀比。

过去的土司住所是更大的碉房,由于规模大、设防标准高,它们基本上就是一座古堡。

四川理县桃坪羌寨　　Taoping village of Qiang people, Lixian county, Sichuan.

四川茂县黑虎羌寨　Heihu village of Qiang people, Maoxian county, Sichuan.

Section 3　Qiang and Tibetan Vernacular Castles

Today's Qiang people mostly live in Aba, located in the valley of the Tibet and Qiang Autonomous Regions. Their neighbors, Tibetan tribes such as Jiarong, should be related to the predecessors of the Qiang people. "Jiarong" means river valley. Since ancient times, there existed much ethnic strife, banditry, and many wild beasts. Therefore, the traditional houses of the locals were all defensive in nature for the safety of the people and livestock. The towers were multi-storied, where the middle floors were used for housing and, storage, and the ground floors housed livestock.

More important defensive buildings are thin, tall towers. Taller than both the Huangcheng Xiangfu-Heshan Tower and the Guangdong towers, its shape is like a chimney. It is used for defense, warehousing, but also similar to the tower in San Gimignano, Italy. They were results of competition and rivalry between families. In the past, the chieftain residences were even greater towers, on a larger scale, with higher security standards, basically resulting in a castle.

四川黑水县的藏寨与相邻的羌寨形式近似，但宗教内容增多，碉房更为高大，但高碉楼少
Tibetan village in Heishui county in Sichuan has similar form to the nearby Qiang village, but with more religions contents. It has more grand tower-house but fewer tall Diaolou.

新旧黑水碉房的对比，新碉房不设防，但保留了色彩鲜艳的出挑式卫生间的传统
Contrasting new and old tower-house in Heishui. New house has no defensive function, but it keeps the tradition of the colourful overhanging toilet.

黑水色尔古藏寨内部，与桃坪羌寨一样，也利用地道增加防御性
The inside of ancient Tibetan village Se'ergu in Heishui county. Just like Qiang village in Taoping, Tibetan village also uses underground passageway to enhance its defensive function.

275

丹巴县梭坡村藏寨中的碉楼群
The Tibetan tower group in Suopo village in Danba county.

下2图：马尔康的卓克基土司官寨的外貌和内部，这种土司城堡公私功能并存
Below: the external and internal views of local Landlord castle in Zhuokeji village in Ma'erkang county. This type of castle has both official and residential usages.